Practicals and Laboratory Work in Geography

Geography Discipline Network (GDN)

Higher Education Funding Council for England
Fund for the Development of Teaching and Learning

Dissemination of Good Teaching, Learning and Assessment Practices in Geography

Aims and Outputs

The project's aim has been to identify and disseminate good practice in the teaching, learning and assessment of geography at undergraduate and taught postgraduate levels in higher education institutions.

Ten guides have been produced covering a range of methods of delivering and assessing teaching and learning:

- Teaching and Learning Issues and Managing Educational Change in Geography
- Lecturing in Geography
- Small-group Teaching in Geography
- Practicals and Laboratory Work in Geography
- Fieldwork and Dissertations in Geography
- Resource-based Learning in Geography
- Teaching and Learning Geography with Information & Communication Technologies
- Transferable Skills and Work-based Learning in Geography
- Assessment in Geography
- Curriculum Design in Geography

A resource database of effective teaching, learning and assessment practice is available on the World Wide Web, http://www.chelt.ac.uk/gdn, which contains national and international contributions. Further examples of effective practice are invited; details regarding the format of contributions are available on the Web pages. Examples should be sent to the Project Director.

Project Team

Lead site: ***Cheltenham & Gloucester College of Higher Education***
Professor Mick Healey; Dr Phil Gravestock; Dr Jacky Birnie; Dr Kris Mason O'Connor
Consortium: ***Lancaster University***
Dr Gordon Clark; Terry Wareham
Middlesex University
Ifan Shepherd; Professor Peter Newby
Nene — University College Northampton
Dr Ian Livingstone; Professor Hugh Matthews; Andrew Castley
Oxford Brookes University
Dr Judy Chance; Professor Alan Jenkins
Roehampton Institute London
Professor Vince Gardiner; Vaneeta D'Andrea; Shân Wareing
University College London
Dr Clive Agnew; Professor Lewis Elton
University of Manchester
Professor Michael Bradford; Catherine O'Connell
University of Plymouth
Dr Brian Chalkley; June Harwood
Advisors: Professor Graham Gibbs (*Open University, Milton Keynes*)
Professor Susan Hanson (*Clark University, USA*)
Dr Iain Hay (*Flinders University, Australia*)
Geoff Robinson (*CTI Centre for Geography, Geology and Meteorology, Leicester*)
Professor David Unwin (*Birkbeck College, London*)
Dr John Wakeford (*Lancaster University*)

Further Information

Professor Mick Healey, Project Director   Tel: +44 (0)1242 543364   Email: mhealey@chelt.ac.uk
Dr Phil Gravestock, Project Officer   Tel: +44 (0)1242 543368   Email: pgstock@chelt.ac.uk
Cheltenham & Gloucester College of Higher Education
Francis Close Hall, Swindon Road, Cheltenham, GL50 4AZ, UK   [Fax: +44 (0)1242 532997]

http://www.chelt.ac.uk/gdn

Practicals and Laboratory Work in Geography

Jacky Birnie and Kristine Mason O'Connor

Cheltenham & Gloucester College of Higher Education

Series edited by Phil Gravestock and Mick Healey
Cheltenham & Gloucester College of Higher Education

Published by:

Geography Discipline Network (GDN)

Cheltenham & Gloucester College of Higher Education

Francis Close Hall

Swindon Road

Cheltenham

Gloucestershire, UK

GL50 4AZ

Practicals and Laboratory Work in Geography

ISBN: 1 86174 027 1 ✓

ISSN: 1 86174 023 9

Typeset by Phil Gravestock

Cover design by Kathryn Sharp

Printed by:

Frontier Print and Design Ltd.

Pickwick House

Chosen View Road

Cheltenham

Gloucestershire, UK

Contents

Editors' preface

This Guide is one of a series of ten produced by the Geography Discipline Network (GDN) as part of a Higher Education Funding Council for England (HEFCE) and Department of Education for Northern Ireland (DENI) Fund for the Development of Teaching and Learning (FDTL) project. The aim of the project is to disseminate good teaching, learning and assessment practices in geography at undergraduate and taught postgraduate levels in higher education institutions.

The Guides have been written primarily for lecturers and instructors of geography and related disciplines in higher education and for educational developers who work with staff and faculty in these disciplines. For a list of the other titles in this series see the information at the beginning of this Guide. Most of the issues discussed are also relevant for teachers in further education and sixth-form colleges in the UK and upper level high school teachers in other countries. A workshop has been designed to go with each of the Guides, except for the first one which provides an overview of the main teaching and learning issues facing geographers and ways of managing educational change. For details of the workshops please contact one of us. The Guides have been designed to be used independently of the workshops.

The GDN Team for this project consists of a group of geography specialists and educational developers from nine old and new universities and colleges (see list at front of Guide). Each Guide has been written by one of the institutional teams, usually consisting of a geographer and an educational developer. The teams planned the outline content of the Guides and these were discussed in two workshops. It was agreed that each Guide would contain an overview of good practice for the particular application, case studies including contact names and addresses, and a guide to references and resources. Moreover it was agreed that they would be written in a user-friendly style and structured so that busy lecturers could dip into them to find information and examples relevant to their needs. Within these guidelines the authors were deliberately given the freedom to develop their Guides in their own way. Each of the Guides was refereed by at least four people, including members of the Advisory Panel.

The enthusiasm of some of the authors meant that some Guides developed a life of their own and the final versions were longer than was first planned. Our view is that the material is of a high quality and that the Guides are improved by the additional content. So we saw no point in asking the authors to make major cuts for the sake of uniformity. Equally it is important that the authors of the other Guides are not criticised for keeping within the original recommended length!

Although the project's focus is primarily about disseminating good practice within the UK a deliberate attempt has been made to include examples from other countries, particularly North America and Australasia, and to write the Guides in a way which is relevant to geography staff and faculty in other countries. Some terms in common use in the UK may not be immediately apparent in other countries. For example, in North America for 'lecturer' read 'instructor' or 'professor'; for 'staff' or 'tutor' read 'faculty'; for 'postgraduate' read 'graduate'; and for 'Head of Department' read 'Department Chair'. A 'dissertation' in the

UK refers to a final year undergraduate piece of independent research work, often thought of as the most significant piece of work the students undertake; we use 'thesis' for the Masters/ PhD level piece of work rather than 'dissertation' which is used in North America.

In addition to the Guides and workshops a database of good practice has been established on the World Wide Web (http://www.chelt.ac.uk/gdn). This is a developing international resource to which you are invited to contribute your own examples of interesting teaching, learning and assessment practices which are potentially transferable to other institutions. The resource database has been selected for *The Scout Report for Social Sciences*, which is funded by the National Science Foundation in the United States, and aims to identify only the best Internet resources in the world. The project's Web pages also provide an index and abstracts for the *Journal of Geography in Higher Education*. The full text of several geography educational papers and books are also included.

Running a consortium project involves a large number of people. We would particularly like to thank our many colleagues who provided details of their teaching, learning and assessment practices, many of which appear in the Guides or on the GDN database. We would also like to thank, the Project Advisers, the FDTL Co-ordinators and HEFCE FDTL staff, the leaders of the other FDTL projects, and the staff at Cheltenham and Gloucester College of Higher Education for all their help and advice. We gratefully acknowledge the support of the Conference of Heads of Geography Departments in Higher Education Institutions, the Royal Geographical Society (with the Institute of British Geographers), the Higher Education Study Group and the *Journal of Geography in Higher Education*. Finally we would like to thank the other members of the Project Team, without them this project would not have been possible. Working with them on this project has been one of the highlights of our professional careers.

Phil Gravestock and Mick Healey

Cheltenham

July 1998

All World Wide Web links quoted in this Guide were checked in July 1998. An up-to-date set of hyperlinks is available on the Geography Discipline Network Web pages at:

http://www.chelt.ac.uk/gdn

About the authors

Jacky Birnie

I am currently lecturing in biogeography and natural resource management at Cheltenham & Gloucester College of Higher Education. Since I arrived here in 1990 I have delivered a wide range of material, including cold climate geomorphology and environmental change, but always with a commitment to maintaining practical and fieldwork experience for the students. Like many lecturers I am aware of the resource constraints on delivery of this experience, and yet absolutely convinced of the value that practical work has to the quality and effectiveness of teaching and learning. My interest in how students learn was sparked off by a PGCE at the School of Education at Oxford University in 1986, where motivating recalcitrant secondary school students provided a very different challenge from that of the post-doctoral research I had been doing at Coventry and Aberdeen. My A level teaching at Andover and Cambridge reinforced my belief that practical work is not just about skills, but about a depth of understanding of the subject. For many students that is the point where geography becomes real.

Kristine Mason O'Connor

My discipline is sociology and in the 1980s I researched and submitted my PhD which was an ethnographic study of the education and employment experiences of young rural women in an English county. I have worked in teacher education and youth work training in Cameroon, Brighton Polytechnic and Cheltenham & Gloucester College of Higher Education (CGCHE). For the past seven years I have combined teaching and research with my role as educational and staff developer at CGCHE.

This role has involved the setting up of programmes of staff development to address the needs of colleagues who are facing increased class size, decreased resources, more diverse student populations, pressures to contribute effectively to the RAE and also generate income. Nationally I have been very involved with the Staff and Educational Development Association and was recently elected as Chair of its Conference Committee.

1 Introduction

1.1 Why provide students with practical work?

This question might be asked by a Head of Department who is concerned about budgets, it might be raised as an issue of teaching quality, or it may be contemplated by a lecturer who is planning or revising a course. Whatever your own concern, this Guide aims to lead you through an exploration of the value of practicals and laboratory work, particularly (but not only) in relation to the teaching and learning of geography at undergraduate level. Case studies and examples are used to share existing ideas and good practice within the discipline, so the Guide provides an opportunity to see what other geographers are doing. It suggests ways to improve the quality of learning that practicals can provide, and outlines new ideas. More than that, it should help you analyse and evaluate your current and future practicals so that learning objectives become explicit.

1.2 What is in this Guide?

The Guide includes analysis (Sections 2 to 5), management (Sections 6 to 8), examples of current practicals (Section 9), alternatives (Section 10) and a Conclusion (Section 11). You could focus on any of these as a starting point, and you may wish to return to the analysis section after reading one of the other sections.

The Guide begins with a section in which the purpose of practical work is examined. This covers a review of research on the effectiveness of laboratory teaching in science, consideration of transferable skills in practical teaching and learning, and checklists to assist you to analyse the purpose of your own practicals. Section 3 examines the concept of 'learning by doing' more closely, to see if it is helpful in explaining the value of practical teaching. Following this, Section 4 asks if the learning outcomes of your practical, including the transferable skills, have been stated explicitly as learning objectives, and this leads on to Section 5 which is a brief examination of assessment options.

In Section 6 we consider the postgraduates who support so much practical work, and the importance of meeting their needs in order to enhance the undergraduates' learning experience. There is also an example of the management of technical support in one department. Section 7 is a brief look at laboratory space and space-planning. Having set the context for the delivery of excellent teaching and learning, Section 8 considers how students might be best prepared for the practical sessions — induction (especially for laboratory work) and health and safety issues are referred to here.

Following this are the details of a set of thirteen case studies of practicals (Section 9). All these have been gathered specifically for the purpose of this Guide by means of interviews. A few have been written-up previously and references to the publications are included. Why these examples? In one (exceptional) case the claim for excellence for a particular practical can be referenced to the Teaching Quality Assessment (TQA) documentation, but almost all of them come from departments rated 'excellent', and all were put forward by their developers as innovations which had been successful in terms of student evaluation and staff satisfaction. The case studies are referred to throughout the Guide, but for some readers this may be the place to start, rather than with the analysis of why we do practicals.

Finally, Section 10 refers to some alternatives to practicals — especially alternative options to laboratory teaching. The Conclusion briefly summarises issues raised in the Guide and proposes that this be addressed by individual geography lecturers or departments as appropriate.

1.3   What is not in this Guide?

Whilst this Guide is 'stand alone' and written to address teaching and learning in labs and practicals within the particular context of geography, there are other more 'generic' key texts which we would recommend (Gibbs *et al.*, 1997; Horobin *et al.*, 1992; Ward *et al.*, 1997)

This Guide does not give emphasis to practical work undertaken in the field or as part of a dissertation; these areas are covered specifically by another Guide in this series (Livingstone *et al.*, 1998). Broadly, a practical session or course, as considered in this Guide, is (or includes) a whole class activity that takes place inside. Practicals may be paper-based, computer-based or laboratory-based. Practicals in both human and physical geography topics are included here. However, there is a bias in this Guide towards laboratory work, which is a form of teaching and learning which is particular to sessions regarded as practicals. Computer-based practical work is reviewed more fully in the Guide by Shepherd & Newby (1998), and activities which could have been included here as paper-based practicals are described as variations on lectures within the Guides by Agnew & Elton (1998) and Healey (1998).

Justifying practicals

*"Is all your lab. work **really** necessary?"*

(Assistant Director, Finance)

Increasing student numbers and diversity, cuts in the unit of resource, 'value for money' audits, modularisation and increasing pressures on academic staff time are key features which require us to rethink the ways we do things in geography and to learn from the experiences of others.

Whilst practicals and laboratory work may be part of the taken-for-granted repertoire of 'what we do in geography teaching', the current climate is generating fundamental questions about what we are doing and why we are doing it.

It is against this background that the ideas and practices in this Guide are proposed.

2.1   What is a practical?

Horobin *et al.* (1992, p.2) offer the following generic view:

"Practical classes, irrespective of subject, depend upon:

1) A material, which may be anything from a document to a stone, from a set of mechanical or electrical components to sets of solutions or people, maps, photographs or machines.

2) A script bearing a task or tasks. This might, for instance, be to construct or run equipment, record in drawing and writing a report on specimens, to analyse text, to prepare a specimen of some kind, perform bodily functions and to record oneself doing it.

3) A protocol which explicitly describes a method or methods relevant to the task.

4) An appropriate environment facilitating the performing of the task. There may be a laboratory with handy benches, balances for weighing, computers, fume hoods, wide tables for maps, safety equipment. In field work the environment provides the material for the class as well as its operational setting. For instance a chalk down, a high street, library, museum or a workplace of some kind."

There are many elements in this broad view which would be familiar to geographers, and it suggests a definition on the basis of what might be **in** a practical, but it does not include **what goes on** during the practical session; we consider that in the next section. We might add to the above that a common characteristic of practicals is their student-centred nature. Students usually generate, collect and analyse data and come to conclusions. Students are involved actively in individual or group tasks. The extent of both academic and support staff involvement is variable, and the format may range from whole class teaching to independent learning.

2.2   ...and what is it for?

"It gives the opportunity to investigate, analyse, test and draw conclusions"

(Geography Student)

Within geography there are two parallel routes for practical work. In physical geography, practicals are often science-based and general research on the role of laboratory teaching in science education is relevant. This is reviewed below. In human geography practicals are less likely to be explicitly scientific (apart from those which train in social science research methods) and may have a more specific focus on transferable skills (see Section 4). However it should be emphasised that, for both areas of the subject, practicals provide a form of learning that might be described as 'active' or 'experiential'. In Section 3 the educational value of this 'learning by doing' is explored further.

2.3   Practicals in laboratories

In the current literature concerned with science education there are a variety of views about the purpose of practicals. Hofstein & Lunetta (1982, cited in Boud *et al.* 1986) emphasise the need to define the goals to which laboratory work could make a special and significant contribution and to capitalise on the uniqueness of this form of teaching. So what is it that practicals, including laboratory work, can achieve that other forms of teaching and learning cannot?

In Woolnough's (1994, p.25) view

"practical activity.. has two functions: to consolidate theoretical understanding and to develop competence at, and confidence in, practical, scientific, problem solving"

According to Boud *et al.* (1986)

"Laboratories are places in which not only skills but also attitudes to scientific inquiry are developed"

Their definition of scientific inquiry includes: observing and measuring, seeing a problem and seeking ways to resolve it, interpreting data and formulating generalisations, building, testing and revising a model.

Sutton (1992, p.72) goes further in abandoning the emphasis on skills, to claim that science lessons (in schools)

"should be the study of systems of meaning which human beings have built up."

However, few geographers would accept that the main function of laboratory work is to assist in the deconstruction of science, as Sutton advocates. Reference to the case studies in this Guide shows that geographers expect laboratory practicals to supply skills training *and* experience in scientific methods.

Horobin *et al.* (1992) remind us that

> *"Practical class work is not research. It is a closed system, a capsule if you like. In the capsule are judiciously chosen material: a narrow, explicitly worded task; a defined time; a prescribed technology. These together determine thus a limited range of outcomes. The situation is contrived to give valuable learning experience through very acutely directed small scale resources."*

And finally McKeachie (1994, p.136) notes that laboratory experience does not always deliver what we intend

> *"While reviews of research on laboratory teaching find that laboratory courses are effective in improving skills in handling apparatus or visual-motor skills, laboratories generally are not very effective in teaching scientific method or problem solving (Shulman & Tamir, 1973; Bligh et al., 1980). Having students design an experiment or analyze data already collected may be a more valuable activity than similar time spent in the laboratory..."*

The above statements are illustrative of some of the many, sometimes contradictory, views of the aims and purposes of practical and laboratory work. The comments also illustrate the gap which may exist between the intentions of these practicals (to train in specific skills or to develop attitudes to scientific enquiry) and the effectiveness of the activities in actually achieving those aims.

In their book 'Labs and Practicals With More Students and Fewer Resources', Gibbs *et al.* (1997, p.3) make the additional point that despite such *analysis* of the aims of laboratory and practical work in science education literature

> *"When we started looking for lists of these aims in course documents...it was a different matter."*

2.4   Clarifying your aims

To support your thinking about the reasons for including practical and laboratory work in your courses, and to share this thinking with the students by way of course documentation, we offer the following three options for consideration:

Firstly, from a review of literature relating to science education at both school and degree level, it is apparent that the principal reasons given for practicals usually fall into one or more of three distinct areas. These are shown in the box below. This analysis provided the basis for clarifying the learning objectives for the case studies in Section 9 when the innovators were interviewed. It is a perception of practicals in geography which underpins this Guide.

Three reasons for practicals in geography

- *Illustrating a theoretical concept*

  In this case the practical is closely integrated into a lecture course, for example dealing with meander generation, and the practical is both demonstrating and reinforcing theory, and illustrating knowledge which might otherwise only be gathered from reading about research. *(cont.)*

- *Providing experience of 'doing science'*

  The practical might involve setting and testing hypotheses, or problem-solving; there might be an element of reflection on — or even questioning of — the scientific process, or on aspects of this such as sampling, or precision and accuracy.

- *Skills training*

  For example, how to design a questionnaire or to record qualitative data, how to produce a soil extract for testing, how to calibrate and use a meter for measurement of dissolved oxygen content, how to identify pollen grains.

Exley & Moore (1993) have noted that practical work in a typical science degree shows a change of skills emphasis with *progression* through the three years. The first year aims were more likely to involve skills training and practice, whereas projects that were tackled in the final year would emphasise problem solving or hypothesis testing.

A second option to clarify your aims is to compare them with those cited by others. Ward *et al.* (1997) list some of the more common aims as being:

Common aims of practical work

- to improve students' understanding of the methods of scientific inquiry through experiments, problem solving exercises and project work

- to develop a number of student skills — measurement, observation, reasoning, problem solving, working in teams, note-taking and presenting work in written and oral form

- to develop professional attitudes to safety and equipment etc

- to develop specialist techniques

- to enthuse students with the subject

- to bridge the gap between theory and practice

- to bring students closer to each other and break down barriers to communication and exchange of views; this may lead to improved seminar and tutorial interaction

- to break down barriers between students and staff

- to develop a respect for the environment

- to expose the complexities of the outside world.

Many of these aims can be identified in the case studies in Section 9, but each example has prioritised one or two objectives, and this has helped to ensure that they are specifically assessed. Otherwise it is all too easy to claim everything, but to be unable to demonstrate that any one outcome has actually been achieved.

Ward *et al.* (1997, p.6) stress the importance of referring back to the specified aims of the laboratory or field work in order to evaluate the students' experience of practical work; they ask 'is laboratory work a vivid and memorable educational experience or is it unthinkable drudgery? Is fieldwork 'camaraderie in the mud' or unreflective learning of bad practice? In other words, is the practical work really achieving what is set up to do?' They emphasise the importance of checking 'the aims and learning outcomes of each course in order to assess the role of practical work in that course.'

A third option is to use a checklist designed by Gibbs *et al.* (1997, p.17) to assist the tutor to engage in a critical assessment of their practical:

Questioning current aims

Aims checklist

- Is it worth doing at all?
- Does current practice achieve stated aims sufficiently well to make it worth continuing?
- Is learning how to learn and tackle unfamiliar situations more important than learning how to tackle identical situations?
- Is it necessary to do it or know how to do it, or is it sufficient to know about it, be familiar with it or appreciate it?
- Are students trained in the skills they are expected to demonstrate?
- Is students' achievement of aims assessed?
- Do students see it or do it often enough to really achieve the aims?
- Can the aim be made explicit and tackled separately rather than remain implicit and buried amongst other aims?
- Is it dull, or can it be made more motivating?
- Which of these features that foster motivation are present?
  - making something that works
  - being oriented towards an end product (such as a car or a device)
  - real world contexts and problems
  - association with modern hardware
  - discovery
  - open-ended problems
  - unpredictability
  - competition, especially between groups
  - cooperation and social interaction
  - novelty
  - challenge, balanced with realistic goals and likelihood of success
  - going beyond the technical content to environmental, social and human issues
  - involvement of 'outsiders' — other years or courses, non-academics.

Use of this checklist might help to focus the practical, and the claims made for it, on those aims it can best achieve. It highlights the need to link assessment to learning objectives, and the importance of fostering motivation amongst the students. Use of the list, and of appropriate student evaluation, can also uncover gaps between intentions and outcomes.

At the moment, whatever experiences a practical may offer (whether illustrating theory, 'doing science' or skills training) they are rarely identified as learning objectives. The innovators reporting on the case studies of Section 9 were clear about their aims and were asked to express them in terms of learning objectives, in other words "By the end of this practical the student will be able to…". This is a process which all of those teaching in higher education are being encouraged to undertake — and clarifying the particular aims that a practical will achieve is a necessary step to writing the learning objectives of the activity.

Interestingly, one way in which geography differs from other sciences is that practicals are rarely set up as demonstrations, with a known outcome, as they might be in chemistry, for example. This pressure for 'the right answer' might lead the students to develop unintended skills as illustrated below.

Confession of a PhD student

"I have engaged in fraud…the plotted graph was nothing but a random splatter of dots. Yet I knew what to do…carefully, I removed some of the dots I did not like. Others I gently nudged towards the middle of the paper…. I was satisfied that my enzyme's activity was properly calcium-dependent…. It was a beautiful graph"

"Why had I cheated so readily? The answer…was that I had been trained to do so"

"throughout my school and undergraduate years, my life had been punctuated by frequent practical classes. These classes had followed a fixed pattern. The demonstrator would explain what we were to do, we would be shown the relevant equipment, and then we would be given some time to do the (so-called) experiment…so-called because there could only be one 'Right' result."

"This 'right' result would be 'divined' from friends, textbooks or journals, and created."

"Our teachers would invariably be thrilled by our expertise, which reflected so well on the design of their practicals, and with honour mutually satisfied, staff and students would pack their books up for the day"

Extract from Terence Kealey, in the Guardian 27.3.97

As the case studies in this Guide show, practicals in geography are more often genuinely exploratory, and the students are gaining insights into environmental research and problem-solving, not into how to fudge the results!

In fact, as noted by Kate Exley (*pers. comm.*, 1997), science practicals in higher education have begun to change in recent years. There has been a move to a problem-centred approach. She sees this as a response to two main influences: perceived gaps in the transferable skills of graduates; and a need to foster greater motivation in the students.

Alongside this, skills in assessment of science practicals have evolved, often now including assessment of specific personal skills such as problem-solving.

In geography we have a greater freedom to design open-ended practicals, since we are rarely demonstrating a scientific principle or law, which would require a predictable outcome. Instead we are providing students with the tools and concepts to conduct enquiries into their environment. Their findings, and their analysis of those findings, may be as valuable as those of a professional environmental consultant, limited only by time and sample numbers. This is a highly motivating process for students — and a very real training in problem solving also.

2.5   Summary

- In a climate of cost-cutting practicals are under threat, and the justification for them needs to be clear.

- The aims are rarely stated in course documentation — tutors need to phrase learning objectives for practical sessions to make the aims explicit.

- Opinions are divided about the effectiveness of conventional laboratory practicals in science as evidence suggests outcomes do not always match intentions.

- The aims of practicals can be distinguished as: illustrating theory, 'doing' science, skills training.

  ⇒ Practicals which illustrate theory may have a 'right' answer.

  ⇒ Practicals which give experience in scientific method will be genuine enquiries.

  ⇒ Practicals which intend to train in skills may need to be repetitive.

3 The value of learning by doing

"Practicals give you insight, you're not just told"

(Geography student)

3.1 Active learning

A common feature of practicals and laboratory work is undoubtedly the 'hands on' element offered to students. 'Active learning' and 'experiential learning' are highly valued by educationalists for the quality, transferability and depth of the learning experience they can provide. Geography lecturers need to look more closely at these terms, which some may dismiss as jargon, because the ideas behind them offer further reasons for practicals. It is surprising what you can find you have been delivering without even knowing it!

> **Active learning:**
>
> *"For students effectively to learn, appreciate, personalise and remember something they must take an active part in learning it. It will not be acquired by passive learning. Meaningful learning is not acquired by treating the mind like an empty pot to be filled; it is more akin to going on a journey"*
>
> (Woolnough, 1994, p.24-5)

This is the argument for getting students involved in doing something — the essence of practical work — not just sitting listening or watching. There is much in the educational literature to support this, and much that has already led to major changes in teaching and learning in schools. There are two main points to note here:

Firstly, the achievement of learning objectives involving understanding is considered to be more efficient where active learning is involved. An example of this might be Case Study 2 where concepts delivered in lectures are cross-referenced to particular practical experiences for the students, and results from the practicals are referred to in lectures. Here the practicals reinforce and develop further the initial conceptual teaching.

Secondly, students are motivated by, and even expect, a high level of active involvement. For example, those that have been following geography A-level courses which include 'decision-making exercises' are used to being presented with a large quantity of information in a variety of formats and having to sort it out in order to present a coherent analysis. Data response questions are used very widely in examinations for 16-19 year olds. The absence of such opportunities for students to engage actively with the material when they commence higher education may be de-motivating.

Activities which have traditionally taken place within practical classes in geography degrees are thus often in line with best educational practice *and* are familiar to increasing numbers of students from their school experience.

Four distinctive features of active learning:

- a search for meaning and understanding
- greater student responsibility for learning
- a concern with skills as well as knowledge
- an approach to the curriculum which looks beyond graduation to wider career and social settings

(Denicolo *et al.* (1992), p.3)

The four features of active learning identified above can be illustrated by practical work in geography. In Case Study 2 the 'meaning and understanding' involves de-mystification of pollen diagrams; in Case Study 3 it is deconstruction of witness accounts of major social change.

Greater responsibility for their own learning is demanded of students in Case Study 1 where project management by the student group is an explicit learning objective, and in Case Study 7 where responsibility is encouraged by role-setting within the student team.

The concern with skills training is sometimes perceived as the only learning objective of practicals, but in geography, in practice, the skills emphasised are likely to be general and transferable e.g. 'Analysing large data sets' in Case Study 6, 'Decision-making' in Case Study 8 and 'Experimental Design' in Case Study 9. Although practical activities may involve very specific techniques, such as measuring the nitrate content of a water sample, any skill involved is a means to an end, and unlikely to be an end in itself. Here geography may differ from the pure sciences and vocational subjects where individual competence in a particular laboratory procedure may indeed be the purpose of the practical, and demonstration of that competence will be part of the assessment.

The fourth claim by Denicolo *et al.* for active learning relates to the wider purposes of higher education, suggesting that active learning, and the development of student responsibility associated with it, is of broad value to the graduate, in a way that sitting in a lecture theatre and taking notes may not be.

3.2   Enquiry learning

"every conscious act done in ignorance of its consequences but with a distinct object of ascertaining what will happen is an act involving enquiry"

(H.E. Armstrong, 1908, cited in Solomon, 1980, p.15)

Enquiry learning is a particular form of active learning, but one that should come with a health warning. Belief in enquiry learning drove the changes in the school science curriculum, where science 'investigations' have taken the place of practice of technical skills, but widespread adoption of this idea also revealed some pitfalls.

In theory, enquiry learning is highly motivating. The promise of discovery leads the student on. Teachers envisaged children learning by free exploration of a set of materials or equipment.

However:

> *"most children wish both to please and be successful and, in the absence of either problem or method, begin to feel insecure"*
>
> *(Solomon, 1980, p.48)*

This doesn't just happen in schools. Many readers will have experienced the corporate negativity that develops in an undergraduate practical class where the structure and purpose is less than clear. A mature student with a tight deadline due to childcare arrangements is likely to be intolerant of uncertainty. The challenge is to offer enough structure for the student to see the point of the exercise, whilst leaving enough freedom for a sense of discovery and individual challenge to remain.

In Case Study 4 final year students are expected to identify their own 'problem' to solve, but are supported by considerable contact with the tutor in the early stages; in contrast Case Study 9 covers similar ground, although at Level II, and experiments are semi-structured. Case Study 10 provides a clear example of fairly open experiential learning, where there is a period of free experimentation in the absence of the tutor. However, the materials and equipment are carefully selected (although including a deliberate 'red herring'), and the concurrent lecture course and reading should help the student groups narrow down the options for their experiments. Nevertheless, this case study remains at the challenging end of the spectrum, and success may depend on reasonably able students and on induction which ensures students appreciate the purpose of the unstructured activity and are willing to see it as a valid challenge.

The box below provides a simple model for examining a proposed practical in terms of its enquiry content. The 'Levels' refer to the degree of openness of the enquiry, not to progress through a degree. In a traditional science practical the enquiry level is low, because the problem is set and the methods are also. At this 'Level 1' the scope for discovery is strictly limited, and it is likely that there is an expected 'right' answer. None of the case studies in this Guide fit into that category, except for the 'Introduction to the Laboratory' course developed for first years at Plymouth (Case Study 5) where the learning objectives are focused on skills training to underpin later work. Final year project work is usually at Level 3 in this model. Students choose the problem to tackle. Practical *classes* tend to be around Level 2, with varying degrees of flexibility in the 'ways and means' available for students to choose. Some also require students to develop their own hypotheses to test, within a broader problem 'envelope'.

An enquiry approach to practical work may be encouraged by the nature of worksheets or guides. The traditional recipe-style protocol is replaced by material which challenges students to solve problems or to set their own hypotheses and devise ways of testing them.

	Problem	Ways and means	Answers
		Levels of enquiry	
Level 1	Given	Given	Open
Level 2	Given	Open	Open
Level 3	Open	Open	Open
Schwab (1962, cited in Solomon, 1980)			

Hegarty (1978, cited in Boud *et al.* 1986) found that use of newly devised laboratory manuals which were more orientated towards enquiry was successful in increasing the time students spent talking about scientific processes in a particular university course.

Many lecturers have discovered that, having developed successful active learning (whether specifically enquiry-based or not), there appears to be no longer a clear (or central) role for themselves. However there *is* a role, depending on the level of support you want to offer. Teaching in this situation means keeping a low profile, but being on hand to assist groups in overcoming obstacles. Being around, rather than disappearing off to mark another set of exam scripts, also means you can observe whether the intended activities are effective. In Case Study 11 this role of the tutor as facilitator is encouraged by a handbook suggesting appropriate activities for staff.

The facilitating tasks might well be passed over to postgraduate demonstrators or support staff (see Section 6), but you would need to ensure that they understood the nature of the enquiry learning you are intending to promote. The innocent helpfulness of other staff can remove every element of discovery that had been carefully planned, as they provide the answers student groups were seeking to discover.

Overall, then, enquiry learning is a major part of practical work in geography, but we would do well to be wary of the extreme positions on the spectrum of possibilities, and to consider how to structure the experience most effectively:

> *"Experiential learning is not the same as 'discovery' learning. Learning by doing is not simply a matter of letting learners loose and hoping that they discover things for themselves in a haphazard way through sudden bursts of inspiration. The nature of the activity may be carefully designed by the teacher and the experience may need to be carefully reviewed and analysed afterwards for learning to take place. A crucial feature of experiential learning is the* **structure** *devised by the teacher within which learning takes place."*
>
> *(Gibbs, 1988, p.14)*

3.3   Deep versus surface learning

In addition to educational concern with active learning, there is an increased awareness of the significance of 'deep' versus 'surface' learning.

Deep and surface approaches to learning

A deep approach is essentially *transforming*, i.e.

- intending to understand material for oneself
- interacting rigorously and critically with content
- relating ideas to previous knowledge and experience
- using organising principles to integrate ideas
- relating evidence to conclusions
- examining the logic of the argument

A surface approach is essentially *reproducing*, i.e.

- intending simply to reproduce parts of the content
- accepting ideas and information passively
- not reflecting on purpose or strategies in learning
- memorising facts and procedures routinely
- failing to recognise guiding principles or patterns

For further discussion of deep and surface learning see, for example, Ramsden (1992).

Building on the arguments for active learning, the provision of practicals as part of a course can be used to assist students in developing a deeper approach to their learning. The encouragement of responsibility for learning, of questioning assumptions and hypotheses, of relating concepts to the interpretation of results, are all aspects of practicals demonstrated by the case studies in this Guide, and by many others. All these contribute to a deep approach to learning and move away from the idea that repetition of routine procedures is the essence of practical work (and the surface learning it would then support).

So active or enquiry learning currently takes place in our practical classes, and is thoroughly supported by research concerned with the effectiveness of teaching and learning. Its continued inclusion can be defended on the grounds outlined above. How can we improve on that experience and ensure that the activities we ask students to engage in really do deliver the experiences claimed for them?

3.4   The importance of reflection

"how does the power and training of the imagination relate to the solid hardware of the laboratory?"

(Solomon, 1980, p.X)

Writers concerned with science education remind us that the learning process we imagine to be confined to the laboratory or practical classroom extends far beyond those walls. It also extends beyond operation of a particular set of apparatus or item of computer software, and involves all the processes of student interaction and reflection which discovery promotes.

"If we are concerned to promote active, student-centred learning, then it follows that students must be given the opportunity to 'process' material for themselves.... If articulating thought and exploring ideas is an important part of learning, then verbal interaction — or dialogue — is a key part of the process"

(Coats, 1991, p.6)

Somewhere in our design of practicals we need to ensure that this opportunity for verbal interaction is promoted and structured effectively. If the design of the room layout (see Section 7) or the organisation of laboratory or computer use (such as bookings for small groups) tends to reduce opportunities for structured dialogue, then we may need to compensate by timetabling sessions outwith specialist facilities specifically for that 'articulation of thought and exploration of ideas'. In Case Study 7 student groups give a preliminary oral presentation of their intended experiment. This contributes 5% to the module assessment. It is sufficient to ensure that students have had to engage intellectually with their subject and are able to explore ideas within a whole class setting prior to embarking on their practical work. Case Study 9 has a timetabled 'follow-up' tutorial with teaching staff to examine the results and reflect on progress made.

"It is not sufficient simply to have an experience in order to learn. Without reflecting upon this experience it may quickly be forgotten or its learning potential lost. It is from the feelings and thought emerging from this reflection that generalizations or concepts can be generated. And it is generalizations which enable new situations to be tackled effectively."

"It is not enough to do, and neither is it enough just to think. Nor is it enough simply to do and think.... Learning from experience must involve links between the doing and the thinking"

(Gibbs, 1988, p.9)

The stages in this process are outlined in Figure 1 by Kolb (adapted by Gibbs, 1988). Consideration of these interlinked stages will contribute to the formulation of a practical or laboratory session which actively promotes student learning.

Figure 1: *Kolb diagram (from Gibbs, 1988)*

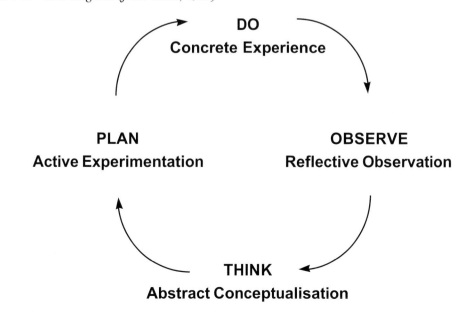

In relation to an existing practical it means, at the very least, identifying when and how students will be enabled both to reflect on their experience and incorporate the learning to inform future activity.

Is there timetabled class time to do this, or is the focus solely on 'the doing'?

3.5   Summary

- Practicals involve active and enquiry learning — this promotes effective learning in line with best educational practice.

- Enquiry learning needs to be well-managed: a given problem and a semi-structured method may be appropriate.

- Practicals involve deep learning where they are concerned with developing understanding and testing hypotheses.

- Effectiveness of the learning can be improved by planning for the process of reflection.

4 Transferable skills development in practicals — being explicit

"What skills do practicals in the laboratory promote?"

(staff developer to student)

"Organisation, time-management, groupwork skills"

(Level 2 Geography student)

Readers are also referred to the Guide in the series "Transferable Skills and Work-based Learning in Geography" by Chalkley & Harwood (1998).

Lecturers often make reference to the importance of skills development in relation to practical work. Whilst technical skills may be acknowledged in objectives, it is also worth making *explicit* which transferable skills are being developed, assessing them and discussing them with the students, not least so that they can confidently refer to these skills when they are preparing their CVs and making job applications.

Denicolo *et al.* (1992, p.4) refer to a range of transferable skills encouraged by the Enterprise in Higher Education (EHE) initiative. You may find it helpful to reflect upon which of these are explicitly being developed in the practicals and laboratory work which your students are undertaking:

Problem-solving skills:
- applying concepts and principles in analysing problems
- producing original or imaginative products or ideas
- using numerical or statistical analysis to solve problems

Initiative and efficiency:
- using initiative, and carrying out one's own ideas
- achieving results within realistic financial and time constraints
- showing greater self confidence
- taking responsibility for one's own development

Interactional skills:
- working co-operatively with others in a group or team
- interpreting and understanding feeling and behaviour
- leading and organising group activity

Communication skills:
- making effective oral presentations in formal situations
- producing effective, well-designed written presentations
- demonstrating computer literacy
- making oneself understood in a foreign language

Why should geographers consider including such skills in our courses?

> *"The cynical view would be that they are the current fashion and this is the way to please institutional managers. Far more important is the need to incorporate those skills which improve the teaching of Geography; enable students to understand better; increase the enjoyment of the subject; and help achieve broader educational aims, such as producing students who are more effective, independent learners."*

(Healey, 1992, p.9)

Case Study 8 involves a variety of practicals delivered as part of a Human Geography module 'Images of the Third World'. The practicals are designed to develop transferable skills of groupwork, decision-making, developing arguments, summarising material and oral presentation. Different practicals have a different focus, although the skills development overlaps. For example, the "game" of keeping and seeking work in week ten of the course, develops interactive skills, whereas problem-solving skills are developed in the population data practical of week three. Case Study 11, again a series of practicals, specifies a particular transferable skill as a learning objective — that of preparing visual, verbal and written material for a variety of audiences. This is developed in the third project of the sequence of four, where group members adopt a role of specialist and put together an oral presentation based on library research.

With the emphasis on transferable skills, it is important not to lose sight of the particular relevance to geography also. If all practicals claim all transferable skills for themselves the purposes of every session will begin to appear the same, and may even be so. Assessment criteria could become identical. Certainly there would be no justification for expenditure on specialist facilities such as computer suites and laboratories if transferable skills, deliverable by less expensive activities, were the only point of practicals. They clearly are not, but sometimes may seem so.

Practicals usually have something unique and subject specific to offer. In Case Study 8, for example, the aims of the whole course were to develop self-questioning and reflection in the students about their own and other perceptions of the 'Third World'. In Case Study 3, similarly, there is an explicit intention to develop student skills in reflection and evaluation of qualitative sources — exposing students to verbatim accounts of nineteenth century issues.

What graduates wish they had

Joanna Bull, at Luton, has been interviewing graduates of the last five years of the School of Geography at Leeds University as part of her PhD research. Of the BSc graduates only a few were using specific techniques in their employment which had been learnt in their geography practical classes, notably those who had entered the water industry, and some involved with weather forecasting. Recent BA graduates were using GIS and artificial intelligence skills in planning, mapping and retail industries. Many graduates felt they were missing IT skills, particularly experience with databases and business software packages, although they all owned up to having tried to avoid IT at all costs at university!

Over the five years there was a trend which seemed to reflect the introduction of enterprise skills into the curriculum. Graduates of five years ago found that they needed better oral skills and abilities in report writing, whereas graduates of 1996 cited fewer of these personal transferable skills and noted rather specialist requirements such as skills in 'Fish Biology' which a geography degree would probably not claim to offer.

(J. Bull, *pers. comm.* 1997)

4.1   Summary

- Most existing practicals in geography develop transferable skills.

- The transferable skills being developed in practicals and laboratory work need to be made explicit in the course documentation.

- The assessment criteria should relate specifically to the skills being developed.

- The acquisition of transferable skills promotes geography graduates' employability.

5 Assessment of practicals and laboratory work

"Assessments for practicals are numbingly predictable — a write up"

(Level 2 geography student)

The topic of assessment is a major issue, (see also the Guide in this series by Bradford & O'Connell, 1998) but what is crucial here is that the assessment of the practical element of a course is planned as part of the initial course design, and is neither 'bolt-on' nor necessarily the traditional science format of every practical being written up using a common style and handed in.

Assessment is the key means of ensuring that the stated learning objectives are explicitly addressed. For example, laboratory practicals assessed by weekly reports might be appropriate where the only learning objective is defined as training in specific laboratory techniques. The quality of the student skills might fairly be assessed by the data they have obtained. If the research skills of analysis and interpretation are also being developed, assessment needs to include those particular aspects by demanding an element of reflection on the results and some evidence of comparison with published work. If transferable skills such as time-management or teamwork are involved, the form of assessment and the criteria involved might be very different; self assessment or peer review may be appropriate.

The key element of reflection, referred to in Section 3, could be encouraged by the keeping of a work diary or learning log, which could contribute to assessment (Habeshaw *et al.*, 1993). Assessing student logs or diaries can be very time consuming for the lecturer. To cut down on lecturer time, a student self-assessment element could be included using, for example, a one page diary/log self-assessment form.

Too much lecturer time can also be devoted to assessing students' laboratory reports.

In the example below it is suggested that the traditional weekly laboratory report might be assessed much more efficiently (and quickly!) if a proforma is developed. The proforma would then be designed to ensure that the student could demonstrate their skills in the specified areas — and only those areas.

Speeding up Assessment with Structured Reports

"Our third year biochemistry course has traditionally had a high content of laboratory work ('practical classes') which have been assessed by making each student produce a report of the work carried out in each class. The classes are of variable length and format...the reports produced by students were often rather long and included repetition of much of the detail provided in the practical class handouts, despite the advice given to students on how they should approach the production of their reports. With increased student numbers on the course, this made marking the reports long and tedious.

The new initiative aimed to streamline the marking process while retaining the value of the practical report as a means of assessing the students' practical performance. Rather than allowing the student to design their own report format, students are now issued with a 'pro forma' report with spaces for them to complete in specific ways. The format of the pro-forma is variable and tailored to each practical, but would typically include spaces for the following types of response:

a)   answers to specific questions posed during the practical

b)   description of specific aspects of their results

c)   presentation of their data in a table, photographic or graph format

d)   reporting overall conclusions

e)   describing any specific problems encountered.

A well designed pro-forma can allow students to present their data in a readily accessible form, clearly indicates to them the amount of detail required for each section to be completed and, together with the practical handout, provides a complete and coherent record of what they have done in the class. This new approach makes it easier and quicker for the students to summarise their achievements in the class."

Stark (1996, p.130)

At Cheltenham & Gloucester College of Higher Education, John Hunt has introduced a semi-structured poster as part of the assessment of a basic first year physical geography module. Some data analysis, graphical exercises and reading are set, and the format of the poster is structured. Students complete graphs in *Excel* and compose relevant text interpreting their graphs and answering some set questions. Students learn how much information can be presented in a poster format, and also have to use word processing and graphics early in their course, with an emphasis on presentation as well as comprehension. Assessment is relatively fast, because the focus is on particular points. Student groups also carry out peer assessment of the posters.

Case Study 2 in this Guide involves a preliminary paper-based practical which is also structured, to assist students on what would otherwise be a steep learning curve for understanding and interpreting pollen diagrams. Completion of this practical prior to writing-up the main piece of coursework based on microscopy means that students have been thoroughly prepared and understand the context of their work.

Another way of saving time and improving the focus of students learning is to involve the student in their own assessment. A recent innovation in assessing a physiology course involved student self-assessment of a basic technique. Students were given the criteria and a grading sheet (such as might have been prepared for postgraduate demonstrators who had been assessing previously) and assessed themselves; they were then assessed by a second student marker to give a comparison. Time management of experiments and writing practical reports were both skills that were successfully assessed this way. The marking process improved the student awareness of the criteria (Kate Exley, *pers. comm.* 1997).

Case Study 7 in this Guide gives an example of students using video as part of their practical assessment, which seems an ideal way of capturing laboratory experiments in geomorphological processes, and incorporating observations of the outcome into an effective presentation for both peer and tutor assessment.

In another example from chemistry, physics and biosciences Harper (1996) developed a new approach to laboratory work which abandoned the traditional practice of tying it to lecture courses. Laboratory work was planned for the whole four years of the undergraduate course which sought to:

- sequence the introduction of higher level tasks
- integrate and apply specialist knowledge and skills
- develop and assess transferable skills
- maximise opportunities for student-centred learning

Harper states that 'tasks of increasing complexity are introduced in a phased manner and the assessment strategy is designed to ensure that competence in all areas is evaluated'.

The range of assessment strategies used included:

- inspection of student lab records
- oral presentation and defence of selected lab records
- assessment of a fully written-up report based on the lab record
- peer assessment of group exercises
- poster presentation of project work
- industrial supervisors contributing to the assessment process
- individual student profiles obtained from the cumulative assessment record

Harper (1996, p.64)

This illustrates that a curriculum may be developed through practical work . It emphasises the importance of progression in what practicals are trying to achieve. The 'whole course' approach is discussed in the Guide on 'Curriculum Design in Geography' (Jenkins, 1998).

There is, however, a danger in detaching practicals from lectures (as Harper envisages) in geography, since every tutor who contributed a case study to this Guide felt that 'demonstrating theoretical concepts' was a major function of their practical sessions. The idea of designing the entire practical experience of students following a particular degree course might be more appropriate for practical work which is clearly focused on technical competencies, but perhaps less useful where practicals are designed to clarify and illustrate concepts communicated in associated lecture courses, as they are usually in geography.

Other methods of assessing practical work are suggested in Brown & Pendlebury (1992).

5.1   Summary

- Appropriate assessment of practicals is a key mechanism for improving the quality of student learning — quite the reverse of being 'numbingly predictable'.

- Assessment of specific practicals needs to be *planned as part of the initial course design.*

- Assessment is the most important means for ensuring that specific learning objectives of laboratory work and practicals are addressed.

- Assessment of laboratory and practical work should promote student reflection.

6 Support

"It was good to have a willing lab. technician to help; we were on our own"

(Level II Geography student)

Delivery of practicals often depends upon postgraduate demonstrators (or graduate teaching assistants) and technical support staff. The undergraduate experience will be tempered by this. The significance of technical support has been highlighted in recent Higher Education Funding Council Reports from the TQA assessments. Technical staff, postgraduate demonstrators and academic staff are all identified as contributing to the quality of teaching and learning in practicals. Good practice involved a favourable staff:student ratio, and staff development which involved all contributors to the student experience.

Teaching quality assessment: extracts from reports

Quality of teaching and technical support

Plymouth (400 single Hons BSc, 120 Joint Hons)

"The ten technical staff play a major and highly valued role in supporting practical work and they, and the administrative staff, have been actively involved in staff development programmes" Oct.94 Report Q18/95, p.5

University College, London (336 undergraduates including 72 joint)

Recommendation to "provide more formal teaching and support for the postgraduates undertaking a teaching role" Nov.94 Report Q66/95, p.5

6.1   The postgraduates

Supporting postgraduates in their teaching role has been on the agenda of many Higher Education institutions in recent years. The following examples featured in the conference 'Training and Using Graduate Teaching Assistants Effectively": Oxford Centre for Staff Development Conference, University of Warwick 13/12/94.

Identifying the needs of Postgraduate demonstrators: The University of Leeds

At the University of Leeds, with over 4000 research students and taught postgraduates (of whom over 700 were on the payroll) the Staff and Department Development Unit (SDDU) conducted a survey of postgraduate needs in relation to teaching responsibilities. Demonstration and marking laboratory work was one of the main concerns of those who replied.

Typical postgraduate problems identified in the survey are listed below:

- Lack of information and guidance from staff on marking coursework and standards

- Lack of appreciation by staff

- Not being paid for preparation and marking time

- Status in relation to undergraduates

- Inadequate feedback from staff and students

- Lack of information on attendance requirements, procedures and resources available

As a consequence of the identified need the SDDU ran a series of workshops during 1994 and about 180 students attended the one on teaching laboratories and practicals.

The Oxford Centre for Staff Development (OCSD) have developed **Resource-based Training** for graduate teaching assistants, piloted by Oxford Brookes University, the University of Nottingham, London Guildhall University and Queen Mary and Westfield College. The materials include a section on 'Labs and field work'.

The Department of Geography & Geology at Cheltenham and Gloucester College of Higher Education (CGCHE) set up an 'In-College' course for postgraduates demonstrators in response their requests for support. This is simply a one-day course, delivered annually, and is in addition to a more extensive 'Support for Teaching' course for all postgraduate research students.

Objectives of the CGCHE 'Preparation for Postgraduate Demonstrating' course:

On completion, postgraduates should feel confident that they understand what is expected from them in the demonstrating role, they should be aware of where they stand in relation to lecturers and undergraduates, what common problems have arisen in the past, some strategies for dealing with them, and where to go for further support.

Further guidance on the role of assistants with laboratory work is provided by Ward *et al.* (1997).

Postgraduates intending to support geography practicals may not always find generic courses in laboratory demonstration helpful since enquiry-based learning may not be assumed in other subject areas. If a course for postgraduates in laboratory demonstration is organised at institutional level, it may lean towards the classic science model: an assumption that technical skills training is the main purpose, and that the undergraduates are moving towards a single 'correct' outcome which will have demonstrated a specific concept. Observation of student behaviour in science classes in higher education certainly found that verification rather than inquiry dominated (Tamir, 1977; Guy, 1982).

As is clear from the case studies included in this Guide, practicals in geography are expected to achieve more than expertise at skills. Demonstrators — and technical support staff — need to be as clear about the learning objectives as you and the undergraduates are, and if the purpose is developing enquiry skills, or problem solving, or project management, the last thing you want is an over-assiduous postgraduate solving all the problems on behalf of the undergraduates.

So, for geography, if you consider postgraduate training offered by the institution is inappropriate, time set aside for briefing demonstrators and other support staff fully about the means and the end of the practical may help them fulfil their role more effectively. A clear diagram about levels of enquiry, such as that from Schwab (1962) which is outlined in Section 3.4 of this Guide, could provide a basis for such discussion. Courses should not take the place of communication between lecturers and postgraduate demonstrators:

> *"Meeting with the lecturer well in advance of the practical session (a couple of weeks, not ten minutes!) in order to be briefed and to be able to ask questions is absolutely vital. It is also important to have the opportunity to meet afterwards to be able to offer feedback and to clarify any ambiguities for another time"*
>
> *(A postgraduate demonstrator)*

6.2   Support staff

> *"To run an extended series of practical classes in a laboratory, you will be dependent upon the technical staff who provide services there. Teaching laboratories are generally in the care of staff whose job it is to keep equipment running, keep the room safe, ensure that nothing is broken or stolen, and to provide immediate support during classes.*
>
> *As well as looking after the room and its services, the technical staff may also be the people who prepare reagents for the class, then keep them fresh and apportion them. The technical staff may also be responsible for storing the mineral samples, rock sections, aerial photographs, maps or other things upon which the class is based, or obtaining the same.*
>
> *In many courses, the technical staff have absolutely crucial roles — the course simply could not run without them. They are supporters and allies to you; they are parent to beleaguered students; they are stern guardians of the laboratory."*
>
> *(Horobin et al. 1992 p.20)*

The authors also remind us, the academic staff, of our responsibilities to technical support staff, who need:

- confirmation of the time and place of each class
- confirmation of student numbers involved, possibly a list of their names
- copies of the protocols a class will use
- a briefing on the practical and details of any changes from the last time the class was run
- a copy of any workbook

Leeds: a case study of excellent organisation of technical support

(Undergraduate student numbers: 620)

The student experience of teaching and learning in the School of Geography at Leeds University is facilitated by a team of support staff. Unusually, students benefit from the work of an **Academic Administrator** whose primary task is the management and co-ordination of all administration relating to undergraduate teaching for the School. This role, which includes managing the process of Course Review (student assessment, peer review and moderating); setting up systems for student records and maintaining records for Teaching Quality Assurance; production of the Student Handbook; advising students; involvement in revisions of the teaching programme and modules; and being proactive in new initiatives in teaching and learning, succeeds in providing a single point of contact for both students and teaching staff in matters related to the administration of delivery.

The development of this position has contributed to the following observations in the TQA report (Nov. 1994, p.5):

"The documentation of various organisational matters concerned with teaching is particularly impressive and permits the efficient audit of the operational details of a well-managed department. The result is a general environment of scholarship and learning which is enhanced by the School's commitment to teaching quality, by extremely good relations between staff and students and is supported by efficient administration".

Excellent support for students at Leeds extends into the laboratories. A **laboratory manager** works with a team of three full time **technicians**. Undergraduates are encouraged to approach technicians directly for advice on project work, and informal 'tutorials' often take place in the technicians' office. The three technicians have developed particular areas of expertise, both in methods and equipment. The laboratory manager takes responsibility for health and safety issues. Technicians manage the booking system which allows efficient use of laboratory facilities, including a considerable amount of student work which is independent of academic staff contact.

Points of contact between academic and technical staff are structured into the course development process. Each practical course is preceded by a meeting of the academic, the technicians and postgraduate demonstrators, and followed up with a 'wash-up' meeting to recommend changes in handouts and protocols. New handouts are written by the lecturer, but modified by the technicians. Proposals for new courses are discussed with the laboratory manager, who would be present at course development meetings.

6.3   Summary

- The effective delivery of practicals often depends on a team of support staff including postgraduate demonstrators and technicians.

- For the undergraduates these support staff are often the main point of contact.

- Postgraduate demonstrators and technicians will benefit from training in teaching and learning, whether offered as courses at institutional level, or in the form of full individual briefing.

- Communication with postgraduate demonstrators and technicians about what the practical is intended to achieve, and how, is vital.

- It is good practice to have an organisational structure which has prior consultation with, and subsequent feedback from, the whole delivery team, built in to course planning and documentation.

7 Laboratory planning

Out of all the comments made in Teaching Quality Assessment reports which relate directly to practical teaching and learning, the great majority were concerned with the physical space in which practicals took place, and the equipment available. A selection of such comments is reproduced in the box on the following page.

Whilst these comments indicate the importance of the space in which teaching and learning takes place, they give little guidance on what is effective or why. Total size of laboratory area is a very crude indication.

How much space?

At Leeds, where excellence was acknowledged, the laboratory floor area totals 466 m². Using the TQA student totals this gives a ratio of 0.75 m² per student. This comprises a main teaching laboratory of 144 m², a large teaching room which supports both 'clean' practicals and meetings (133 m²) and two smaller laboratories (115 m² and 75 m²) for projects and groupwork, which are shared with Masters students and postgraduate researchers. These latter rooms are less suited to class teaching as much because of the amount of equipment they house as because of their smaller floor area.

Leeds makes good use of this purpose-built suite, which allows different undergraduate activities to be timetabled into different areas. Much work is of an independent nature (although supported by technicians) and student groups can book into one of these laboratories at almost any time of the day or week. As undergraduate demand tends to occur in blocks of time, technicians block research access also. The needs of large class teaching and small, and flexible, group work, are thus both met. The only cloud on the Leeds horizon was the possibility of the introduction of space-charging to departments. Where this happens, the pressure will be for multipurpose use of these designated spaces.

(*pers. comm.* Mrs. A. Kelly, Laboratory Manager, 1997)

Room designs express the assumptions of designers about how learning will occur. Once established, rooms can shape the activities that take place within them and "create a culture in which people think there is no other way to work" (Sutton, 1992, p.83). Many geography departments rely on access to traditional science laboratories for their practical work. Some departments, despite having had the opportunity to fit out their own laboratories in recent years, have re-created traditional laboratory spaces. It is worth questioning whether this is the most appropriate alternative for effective teaching and learning of what it is that geographers value in their practical classes.

7.1   Why conventional science laboratories?

Laboratory classes in science subjects at university are relatively young. Sutton (1992) records that practical classes arose from student demand and the demand was driven by vocational need. They were not established in all science subjects until late in the nineteenth century, and, prior to that, knowledge was imparted by means of lecture-demonstrations. The costs of the demonstrations were borne by the professor, whose income depended on the number of students choosing to attend the lecture. It was a model that did not encourage hands-on experience, not least because of the loss of mystique and status involved. However, once teachers outside the universities began to supply the individual practical experience students wanted, and once it was recognised that researchers required training, institutions began to see the need for practicals.

Those early practicals emphasised procedure. They were a form of advanced apprenticeship training, "an induction into some of the procedures of the discipline" (Sutton, 1992, p.85). The rooms in which this induction was to take place were designed with this in mind, modelled on the laboratories of working scientists. Laboratory manuals conducted large numbers of students through the correct and identical processes, and, for many of us, school science classes were a continuing echo of that emphasis. The rooms were designed for apprenticeship, and not for teaching and learning as we now understand it.

School science, however, has changed. Epitomised in the Nuffield Science courses, secondary science teaching has shifted emphasis onto research skills, student investigative projects, discussion about where ideas come from, and reflection on the nature of science. New school science laboratories reflect these aims — a need to cater for face-to-face discussion, for groupwork, poster displays, oral presentations and individualised experimentation. Tables have replaced standard science benches, and the teacher moves around from group to group, facilitating the practical, not delivering from the front.

Teaching quality assessments: extracts from reports
(statistical terminology reflects differences in the published reports)

Accommodation

The majority of comments on practical work in the TQA reports were related to accommodation, suggesting that this is of prime importance to teaching quality. Laboratory facilities were needed to provide 'scientific' courses. A lack of access to practical training in technical skills was seen to limit the dissertation topics in physical geography.

Birmingham (620 undergraduates)

Noted there were four small laboratories available for undergraduate work in physical geography, but that a further development would soon be completed. Dec.94 Report Q48/95

Cambridge (290 FT)

Laboratory space for physical geography is marginal and the equipment outdated (p.5)

Recommendation "to review and improve laboratory provision which in quality, quantity and safety is barely adequate for the essential needs of physical geography courses"

"Student attainment is noticeably higher in human geography than in physical geography, which is less favoured in the number of courses, in laboratory provision and in the choice of dissertation" (p.6) Jan.95 Report Q132/95

Cheltenham & Gloucester College of Higher Education (232 FT)

"The assessors were particularly impressed by the laboratory provision..." (p.5). Noting "The attractive accommodation and appropriate physical resources for teaching and learning including ... laboratories." (p.6).

Durham (365 undergraduates)

"The laboratories...are not suitable for multiple research and teaching use, and modern scientific teaching is hampered by lack of space" (p.5)

"There are problems with the teaching of physical geography in which significant subject areas are virtually omitted" noted that this limits the dissertations (p.6). Feb.95 Report Q185/95

Kings College, London (237 undergraduates):

Recommended "the urgent continuation of the teaching and laboratory building improvements programme, as some accommodation provision...is barely acceptable" Oct.94 Report Q9/96 p.7

Lancaster (263 undergraduates)

"The provision of a 400 m^2 laboratory for physical geography is an excellent recent addition to the departmental facilities" (p.5) Jan.95 Report Q3/96

Liverpool Institute of Higher Education (600 including Environmental Studies)

"...has a substantial new laboratory block for physical geography and environmental studies — offering a very attractive and well-equipped working environment." Jan.95 Report Q72/95

Manchester (410 undergraduates).

Noted that increased student numbers, together with the ageing profile of the equipment, have reached the point where refurbishment of the laboratories is becoming a necessity. Jan. 95 Report Q143/95 p.5

Oxford (300 students)

"Although all the laboratories are well equipped, they are small and currently available only for research and small option teaching. The expansion of these facilities would improve the resources to support this subject". Feb.95 Report Q166/95

Oxford Brookes (114 FTE)

"The provision of physical geography resources did not fully reflect the scientific nature of some of the geography modules taught, particularly with regard to the need for analytical laboratory facilities"

Noted that this constrained the opportunities for geography dissertation and project work.

There was "insufficient access to teaching laboratory space for training geography students in environmental monitoring and analysis" Jan.95 Report Q133/95 p.5.

Sheffield (550 including joints)

Noted excellent teaching sessions which included laboratory practicals.

Teaching and research laboratories were well resourced. Feb.95 Report Q162/95

Within higher education, and particularly in laboratory-based practical classes in geography, the room and the structure of its fittings may emphasise technical skills training even though our stated learning objectives are something else. There are alternatives. At Cheltenham and Gloucester College of Higher Education the Department of Geography & Geology moved into newly-fitted accommodation in 1994. After consultation with all teaching and support staff whose undergraduate courses involved an element of laboratory-based work, a physical geography laboratory was designed which broke away from the conventional rows of benching, and visually-intrusive water taps (Figure 2).

Figure 2:   *Sketch layout of laboratory at Cheltenham & Gloucester College of Higher Education*

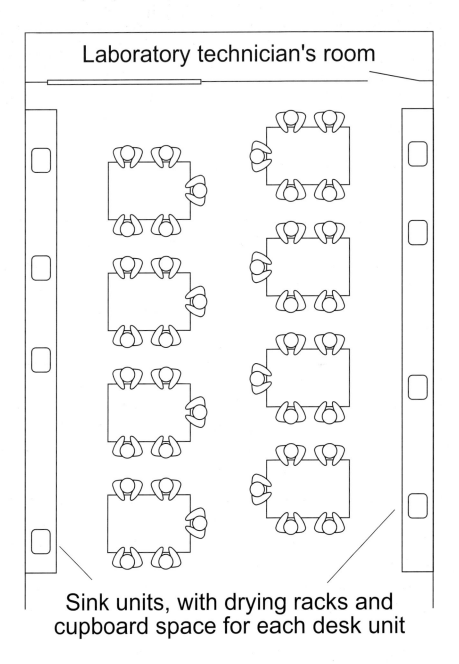

Laboratory technician's room

Sink units, with drying racks and cupboard space for each desk unit

Design for a new teaching laboratory
Cheltenham and Gloucester College of Higher Education

The laboratory was designed around eight workbases (Figure 2). Each student group works around a simple, fixed table which is free-standing, whilst benches with standard services (water, gas and vacuum) and cupboards and drawers of glassware and other standard equipment occupy the sides of the room. Each group has access to its own sink, drainer, services and equipment. The group is encouraged to be self-sufficient at its base, and to clear everything away at the end of the practical.

Interaction within the group is maximized, seated on stools or standing around their table. Cross-laboratory sorties are minimized. Lecturing from the front is possible, since the room is visually uncluttered. Videos which demonstrate a range of practical techniques are available in a small adjacent area for independent or group viewing.

7.2 Learning objectives can save money

Laboratories are expensive, but laboratories for geographers can be quite low-tech. If rooms for practical work are designed around meeting the sorts of learning objectives that have been outlined in Sections 2, 3 and 4 of this Guide, rather than based on conventional science laboratories, the demands are considerably less in terms of specialist furniture and fittings. The arrangement of the laboratory at Cheltenham and Gloucester College of Higher Education (see above) means that the room is also used by human geographers who want to work more flexibly with groups on paper-based practicals.

Geography undergraduates do not need hazardous chemicals, ceramic sinks, or gas taps, but they do need basic stainless steel sinks and drainers and large-capacity sediment traps. They need electricity, and vacuum pumps are helpful for filtering sediments. They need access to drying ovens and furnaces, and perhaps a fume cupboard — but these do not have to be in the teaching room. Glassware and equipment requirements are simple for students whose main concern is with sediment, soil and water analyses.

Where the learning objectives of the practical experience include concern with setting and testing hypotheses, problem solving, groupwork and demonstrating concepts there may be little justification for undergraduate access to complex and expensive equipment, but much requirement for discussion and reflection. An awareness of the existence of such equipment and its applications — such as an Atomic Absorption Spectrophotometer (AAS) and its use for analyses of lead levels in tapwater — may be met by a single demonstration or a video. It is hard to make a case that every geography BSc undergraduate actually needs to use it. In the future, undergraduates may be able to access demonstrations of state-of-the-art equipment on the Internet.

7.3   Summary

- The physical space for practicals is essential, and has been the prime focus of attention on practicals in geography by Teaching Quality Assessment teams in the UK.

- A ratio of 0.75 m^2 of dedicated laboratory floor space per geography undergraduate is a guideline.

- The planning and arrangement of laboratory furniture has an effect on the nature of the teaching and learning: what was appropriate for skills training in the nineteenth century is not likely to be appropriate now.

- Laboratory planning for geography should be based on the learning objectives of the practical curriculum and this may be a more economic option than adopting the traditional science model.

8 Student induction to laboratory and practical work

"Yesterday I saw all this equipment and I felt 'oh no I don't know what I'm doing'"

(Level II geography student)

8.1 Student induction

Widening access, student diversity — in particular the increase of students from 'non-traditional' backgrounds — and modularisation of degree courses present the geography lecturer with a group of students with a wide range of previous experience. This can pose a challenge in terms of the effective use of laboratories for teaching and learning.

Geography students studying at the same level within a modular scheme indicated that even within one institution there was a wide range of experience in terms of being prepared for work in the laboratory:

"Laboratory work was explained beforehand, but not in detail — perhaps a detailed procedure handout for methods would have been useful to read the week before"

"Lectures beforehand gave us an idea about what we may find out in the lab. However no instructions were given explaining how to perform the tests etc."

"The preparation is generally quite rushed, it is given while in the lab because time is extremely limited and consequently all the work has to get done. You have to rush."

"Written and orally presented instructions/guidelines, including health and safety in the lab"

"You come here and they say this is what you do and you just do it"

"I'm doing Human Geography and I'm quite blind to all this."

In view of the diversity of student background, particularly on modular geography degree courses, it would seem advisable to incorporate student induction to laboratory work. Those considered to have appropriate backgrounds could be required to present evidence of their knowledge and skills.

The 'circus'

In Case Study 1 student groups are introduced to the appropriate laboratory equipment and techniques for their water quality analyses when they book a six-hour laboratory visit. During that time academic and support staff are mustered to demonstrate the equipment to the student groups, as they move around the stations in the laboratory. The main teaching laboratory is devoted entirely to this class in the week in which induction takes place. Following that, students are able to work in the laboratory with much less support, and after the investment of time at the beginning the staff requirements are much reduced as the groups work through their projects.

Gibbs *et al.* (1997, p.21-22) suggest that a library of instructional videos for various experimental instruments and rigs is made available to all students. They are asked to become familiar with these instruments on an 'open lab' basis and are then tested by multiple choice questions or brief observation by a technician.

Another example from the same source is a variation on the 'circus' with peer tutoring of equipment use. In the first week each student group is given a piece of equipment with which to become familiar (and possibly some exercises or experiments which will help this process of familiarization). Each group become the 'experts' on their piece of equipment for future weeks, briefing the next group who need to use it and helping them to overcome any problems they encounter. Groups rotate round equipment.

Workbook

Case Study 5 is designed as a Level 1 induction to the laboratory, and the practicals are supported by a very comprehensive workbook. Within the book are not only the step-by-step instructions for a wide variety of simple practical exercises, but also COSHH (Control of Substances Hazardous to Health) regulations and risk assessments where appropriate. Results which students are expected to obtain are partially structured for them, with tables and graphs within the handbook to complete, in addition to communal results sheets in the practical class which are then available to the whole class on the file server.

From interviews with geography students it is clear that there does exist considerable anxiety about equipment usage and about the unfamiliarity of a laboratory environment. Similar concerns are experienced in relation to computers. Many students may focus so much on that one aspect of the practical — not knowing what to do and being concerned about getting it 'right' — that they don't perceive the other functions the exercise may have. Some sort of familiarisation process, whether of laboratory apparatus or computer keyboards, in a supportive environment might be very beneficial. As laboratories and associated support staff are considered to be expensive, and as student groups often have to be split into several sub-groups to allow them access to the laboratory, there is a tendency to push them through with some haste, expecting effective production of results to emerge quickly. Given the variability of prior experience, we should pay much more attention to student induction.

8.2   Health and safety

Health and safety should inform and underpin all good practice in practical and laboratory work. You are recommended to read the Health and Safety Executive publication related to laboratory work (HSE, 1990). The Health and Safety Executive Information line is 0541 545500, WWW site: http://www.open.gov.uk/hse/hsehome.htm.

Safety in Laboratories

Institutions and departments should have clear, readily accessible and known guidance policies and procedures for health and safety. An example related specifically to geography is contained in the first section of the University of Plymouth's 'Laboratory Skills' Guide produced by the Department of Geographical Sciences:

"Staff and students at the University of Plymouth are subject to the provisions of the Health and Safety at work legislation, including the recent COSHH Regulations. In summary this legislation imposes a duty on all staff and students to carry out laboratory work as safely as possible."

The booklet proceeds to address issues of COSHH, dress in the laboratory, general behaviour and undertaking experiments. An example is then provided of a COSHH assessment and the international hazard warning symbols.

As well as providing key information about health and safety the issue itself could be integrated into the curriculum so that it is not seen as a tiresome 'bolt on' extra.

Integrating safety issues into the course

A strategic approach is to introduce and involve issues of safety in all aspects of your courses. Deciding to do this will probably influence the way you prepare course documentation, as well as the way you run the practical activities themselves and, finally, the assessment process.

Horobin *et al.* (1992, p.60)

8.3   Summary

- Geography undergraduates, particularly on modular degree programmes, commence their studies with a very wide range of prior experience related to laboratory and practical work.

- Student induction to laboratory and practicals needs to be incorporated into the geography course design.

- There is a wide range of mechanisms for student induction from 'hands on' in small groups to individual instructional videos.

- The content of student induction programmes should include not only equipment use but also the purposes of practical and laboratory work.

- An understanding of, and adherence to, health and safety issues is essential for both students and staff in geography practical and laboratory work.

Case studies

9.1   Introduction

This section includes details of case studies of current practicals in geography undergraduate courses in the UK. The purpose of this part of the Guide is to illustrate a range of practice in geography departments which is to be found in higher education in the UK at the present time, and to provide real examples for further analysis and evaluation by the procedures described in Section 1.

You may prefer to apply that analysis to your own current teaching. Or you may be interested in the case studies as a source of ideas from colleagues in relation to particular subject areas.

The format of the presentation of the case studies reflects their source and their role in this Guide. The information about each case has been obtained by interview with the innovator. *Their* description of the session is summarised under a series of common headings. At the foot of each case study, in a 'Comments' box, we have drawn attention to aspects of the practical session(s) which we believe link particularly with themes in the rest of the Guide. For further information about each case study the reader is referred to relevant publications, or to the innovator themselves.

9.1.1   Themes addressed by particular case studies: quick guide...

	Case Study
Student induction	*1, 2, 5, 9*
Time-management	*1*
Qualitative research	*3, 8*
Motivation	*4, 13*
Large classes	*5, 11*
Links to 'industry'	*6, 12*
Use of video	*7*
Using heavy machinery	*7*
Diversity	*8*
Reflection	*9, 13*
Enquiry learning	*10*
Issues-based	*11, 13*
Low cost options	*4, 10, 12*

Case study 1: Water quality problems in Leeds

Innovator:	Pauline Kneale, Leeds
Practical:	Group projects involving laboratory testing of water quality
Level and numbers:	Level 2, 92 to 106 students
Preparation:	No pre-requisite, 2 introductory lectures and 2 workshops (teamwork, and poster design)

Context:

Students are set the task of developing and testing hypotheses concerning water quality of the tributaries of the River Aire. In groups they have to then carry out field sampling, laboratory testing, data analysis, and complete a poster report and oral presentation. The whole unit lasts 5.5 weeks. The laboratory practical work is therefore part of a logical sequence of activities designed for environmental problem solving. It uses samples collected by the students. The outcome is not designed by staff.

Aims:

- Link to concepts delivered in *Urban Hydrology* and *Resource Management* modules

- Experience of 'doing science' in the sense of hypothesis setting and testing

- Experience of project management

- Some skills training.

Techniques included:

- Measurements of nitrate, phosphate, colour, hardness, calcium, magnesium, suspended sediment; and others 'after discussion'.

Management of laboratory access:

Laboratory induction is part of the initial six-day briefing period. Staff, three technical and one academic, set up demonstrations of all available methods around the laboratory. In a six-hour block student groups attend and move around the stations, learning how to use the equipment. They make decisions about the analyses appropriate to their hypotheses, and then book laboratory time, field equipment (time of collection and return), insurance cover, and complete a safety review, through the technicians, according to their project time management plan. For the week in which induction is taking place the main laboratory is entirely devoted to this course.

(Support staff arrangements at Leeds are described in Section 6.2)

Assessment:

A group report of the field and laboratory work, which includes a diary and reflection on success with proposals for further work. A group poster which is explained in a brief oral presentation. Students assign the mark allocation within their own group.

Reference:

Kneale P. (1996) Organizing student-centred group fieldwork and presentations, *Journal of Geography in Higher Education*, 20(1), pp.65-74.

Comment:

The teamwork workshop, described more fully in the reference above, is unusual in preparing students for effective groupwork. It is probably particularly important, given the amount of choice student groups have, in what is fairly open enquiry learning (see Section 3). Students are highly motivated by the amount of freedom and responsibility they have. Time-management and groupwork are transferable skills which are explicitly developed.

Case study 2:   Environmental change in Shetland

Innovator: Lisa Dumayne-Peaty, Birmingham University

Practical: Individual student work on palaeoenvironmental reconstruction

Level and numbers: Level 2, 120 students

Preparation: No pre-requisite, 1 introductory lecture.

Context:

Students are provided with a preliminary exercise (on paper) in which they are taken, step-by-step, through the procedure for zonation of an existing pollen diagram from Lewis, including addition of dates, comparison with published isopollen maps, and interpretation of the vegetation record. This illustrates concepts referred to in concurrent lectures (and means that techniques of interpretation are not left until after their counting in the main project). Pollen taphonomy, dispersal, problems of interpreting pine pollen curves included. Actual results from the practicals are referred to in the lectures. In microscope laboratories, with assistance of technician or postgraduate demonstrator, students identify and count pollen and chironomid slides prepared for them. Identification is principally from illustrations in student practical books, not from type material. Students then apply the interpretation skills to their own data.

Aims:

- Skills training — introducing techniques of palaeoecology, training in identification techniques
- Link to concepts delivered alongside in lectures

Students do not set hypotheses, but course is designed to emphasise discrepancies between theory and reality ('Science is messy'): based on site records which contradict isopoll maps. Students have to choose what to accept and what to reject.

Learning objectives:

When students have completed this practical they will:

1)   be able to identify basic pollen and chironomid types

2)   be able to interpret palaeoecological data

3)   be able to relate theoretical concepts to real data

4)   understand the processes involved in Holocene environmental change

Techniques included:

- Pollen diagram zonation (paper-based practical)

- Pollen identification and counting

- Chironomid identification and counting

Management of laboratory access:

The microscope laboratory only holds 12 students, so during the 6 weeks of the course there are eight repeats of 2 hour sessions, plus one day on which the laboratory is open to anyone. Timetabling avoids clashes with other courses.

Assessment:

The first 25% of assessment weighting is on the paper-based practical which introduces students to a pollen diagram and the processes of interpretation. The main research report is 3000 words, detailing and interpreting results from the laboratory practical. The structure is set. Pollen and chironomid diagrams created from identification and counting in laboratory practicals are included in the write-up.

Comment:

This practical set represents a careful balance between structure provided and independent student activity, which has evolved over time. Palaeoecological techniques are hard for students to learn because of the dependence on skills of microscope use and then microfossil identification. Other courses have run into problems with maintaining student motivation sufficiently to reach a worthwhile output. The provision of the initial paper-based practical which gets to grips with the point of the whole exercise quickly overcomes that problem.

Case study 3: Crofting — Voices from the past

Innovator: Iain Robertson and Charlie Withers[1]
Cheltenham & Gloucester College of Higher Education
[1]now at the University of Edinburgh

Practical: Individual student work, paper-based practical on original sources

Level and numbers: Level 2, 25 students

Preparation: No pre-requisite, 1 introductory lecture on qualitative methods and historical context

Context:

This is a single practical session, one of five in a course entitled 'Reconstructing Historial Communities'. Students are provided with a workpack which comprises 10 hours of independent work over a two week period. The nature of that work is primarily reflective. Extracts from the Napier Commission report are reproduced in full, and comprise the verbatim evidence of two people in contrasting positions: a Skye crofter and a famous

Highland agitator. Contemporary (nineteenth century) illustrations, and photographs of evictions are included in the workpack. The students complete an assessed exercise which calls on them to develop skills of reflection and evaluation of qualitative sources. Reading lists are supplied.

Aims:

- Skills training — in qualitative research methods

- Link to concepts delivered alongside in lectures

- Broaden student thinking

Learning objectives:

1) To challenge students' preconceptions and distrust of qualitative research. Too easily and often, students equate subjectivity with 'bias' and 'inaccuracy'. Subjective opinion is contrasted negatively with supposedly objective data sets and condemned for a 'lack of objectivity'.

2) When students have completed this module they will have gained an appreciation of the strength of qualitative research, developed a more reflective attitude towards the gathering and interpretation of data, gained a deeper understanding of historical-geographical change in the Scottish Highlands.

Techniques included:

- Reading and reflecting

    - from the introduction to the Resource Pack:

    "the tyranny of data sets and statistical analysis is replaced with the analysis and interpretation of emotions; opinions; interpretations; observations; memories; beliefs; ideologies — those of the researcher as much as the researched"

Assessment:

There are two written exercises set for this practical, both of which expect students to have read and reflected on the qualitative material provided, and also carried out background reading on qualitative methodology. However, students choose which units of this course to write up and submit for assessment, and not all students carry all the practical activities through.

> *Comment:*
>
> The distinctiveness of this practical is its involvement with emotions and responses, and the initiation of students into qualitative research on historical sources. Students find it hard due to the lack of structure, and the requirement to draw on their own resources of reflection and intellectual activity. However, some students perform extremely well in this area, where others will prefer more familiar 'factual' practical exercises. It develops skills other practicals may not reach!

Case study 4: Experiments in geomorphology

Innovator: Alastair Gemmell, Aberdeen University

Practical: Individual student projects

Level and numbers: Final year, over 20 students but only 5 chose this rather than a library-based project

Preparation: No pre-requisite other than successful completion of second year physical geography course. Students have very mixed backgrounds.

Context:

These are laboratory-based student projects taking place in the latter half of a unit in *Applied and Engineering Geomorphology*. On the basis of the concepts already delivered in lectures, and wider reading, individual students are encouraged to design and carry out an activity which involves generating and analysing data relating to a particular geomorphic process. Successful examples have included the following: freeze/thaw weathering of building material or different rock types; soil erosion under different stages of a (cress) crop; effect of bridge piers on river channel scour; strength required to move ice over different grades of sediment; modelling action of boulders in basal ice — production of conchoidal fractures in bedrock (Smith, 1984).

Aims:

- Skills training — principally in managing an investigative project, including designing, conceptualising, implementing and running the experiment, and assessing the shortcomings

- Practical problem-solving

- 'Doing science' in the sense of setting hypotheses and collecting data to test them

- Link to concepts delivered previously in lectures

Learning objectives:

When students have completed the practical exercise they will be able to:

1) conceive, design and implement a research experiment based upon laboratory work

2) generate appropriate hypotheses for testing through the experiment

3) analyse and critically assess data generated by the experiment in terms of the hypotheses

4) better understand the bases and limitations of existing theory in at least one area of geomorphology

5) write a well-structured and coherent report of an experiment

Techniques included:

- Specific to project, but students encouraged to design and use 'shoestring' apparatus, often simulating more complex equipment described in research publications.

Management of laboratory access:

Student time in the laboratory is managed by the laboratory technician, there is not an access problem

Assessment:

Project write-up is assessed as 33% of the course, the remainder is an examination.

> *Comment:*
>
> Practical work delivered by means of individual projects is highly motivating for the students who choose to undertake it — it is disappointing that the majority select a library-based project instead, which is perceived to be less challenging. The aspect of the practical that is particularly noteworthy is the simplification of complex, research-level, apparatus (by the students) to simple materials, which yield highly successful experimental results. The results, in turn, stimulate much background research in relevant literature.

Case study 5:   Introduction to the laboratory

Innovator: Ruth Weaver, University of Plymouth

Practical: Introductory course in laboratory skills

Level and numbers: Level 1, 170 students

Preparation: No pre-requisite, students have both Arts and Science backgrounds

Context:

This is an introduction to laboratory skills, comprising half of a Level 1 module. The other half is on graphics skills and is concurrent. There are six sessions, each of one hour. A selection of techniques and equipment are covered, including soil, water, sediment and microfossil analyses. It is compulsory for BSc geography students and ensures they cannot graduate without at least some laboratory skills. It underpins Level II modules. Students work in groups and are provided with a handbook.

Aims:

- Skills training
- Elements of illustrating concepts and 'doing science'.

Learning objectives (in order of importance):

Students should be able to:

1) use basic laboratory equipment correctly and with confidence

2) understand the need for careful, accurate and precise recording of results

3) recognise scope and causes of experimental error

4) relate text book descriptions of physical properties to measurements they have made in the laboratory

Techniques included:

- Soil colour, moisture, organic matter, texture, mineral content

- Dry sieving for particle size

- Chloride concentration of water using spectrophotometers

- Use of light microscopes (pollen) and use of identification keys

Management of laboratory access:

Teaching laboratory (shared with biology) accommodates a maximum of 35 students; so the unit is run six times, for 30 students at a time. Booking sufficient laboratory time and technician support time is not always easy. Technical support is excellent. Apparatus and samples are set up beforehand (one hour), at least one technician and one academic (lecturer or demonstrator) support the practical, technician clears away in about 20 mins. Running costs (materials and equipment) are not an issue because very little is used. Future management will depend on demonstrator and support staff availability.

Assessment:

Students put their soil analysis results on a communal results sheet which is then held on the geography fileserver. For the assignment they use simple descriptive statistics and are asked to provide a brief discussion also. For assessment of the particle size practicals their answers to structured questions (completion of tables and graphs) are handed in.

Comment:

This practical series is a useful example of: a) providing induction for students, including health and safety issues (COSHH regulations and risk assessments are included in the excellent handbook which supports these practicals) and b) overcoming the problem of delivering practical laboratory experience to a very large class — using multiple runs and technician support — and managing the results and assessment imaginatively with the communal record sheets, and results accessed on a fileserver.

Case study 6: Bracken invasion

Innovator: Ruth Weaver, University of Plymouth

Practical: Mapping moorland vegetation

Level and numbers: Level 2, 35 students

Preparation: Most students will have taken Level 1 biogeography and a general practical course at Level 1 which introduces air photos and satellite images

Context:

This is a set of three practicals, each of two hours, with much staff support. They are delivered in the latter half of a module on *Vegetation Dynamics and Management* which is intended to show how knowledge of vegetation dynamics (specifically competition and succession) can help in the practical management of plant communities. The first exercise is to map moorland vegetation from air photos and satellite images; the second involves data analysis of the results of a bracken-encroachment experiment on the North York Moors; the third, also data based, uses the results of a long-term experiment on Dartmoor on the effects of fertilising and grazing on heather. Students work individually, but consult. The conceptual grounding is given in lectures beforehand, what the practicals do is provide the students with real data and real problems. The first exercise promotes debate about the decisions that have to be taken in producing a map, the later exercises provide experience in dealing with large amounts of data, and solving problems. They are quite challenging.

Aims:

- Skills training — map production and analysing large data sets

- Illustrating concepts

- Experience in real management problems

Learning objectives:

Students should be able to:

1) grasp difficulties of drawing conclusions from complex, real-life datasets with several interacting variables

2) relate concepts of vegetation dynamics to practical examples

3) interpret various kinds of aerial/satellite imagery and understand their limitations

Techniques included:

- Interpretation of aerial photographs, OS maps, satellite images

- Analysis and interpretation of large data sets in the form of TWINSPAN tables and other summary statistics

Assessment:

One assignment is set on each practical and students attempt all three, but only complete and hand in one. Each consists of the results of the practical work plus a written discussion based on a series of points of 'suggested analysis' plus background reading.

Comment:

This example is distinctive in its use of data of professional quality relating to real environmental problems. This brings students in contact with data sources of a realistic size and complexity. Students find this challenging, and need support, but benefit from the experience of 'real world' environmental consultancy.

Case study 7:   Experiments in soil erosion

Innovator:	Martin Haigh, Oxford Brookes University
Practical:	Group projects with a large tipping flume
Level and numbers:	Level 2 or 3, 55 students on module, fewer on this project
Preparation:	No pre-requisite, 1 introductory lecture and demonstration

Context:

As part of a module entitled *Soil Conservation* students work in self-created teams to explore the controls of soil erosion (Haigh & Kilmartin, 1987). They use a large laboratory soil tray as a test bed for conducting their own investigations into the ways in which slope angle, surface water run off, subsurface water, sediment size, organic matter, surface cover, surface roughness, rilling and/or simulated rainfall affect sediment mobilisation. They may also design experiments that test the effectiveness of different erosion control technologies, such as transverse barriers, mulches or terraces. Lectures in soil erosion processes and technologies run alongside the practical project. The project is supported by a 40 page workbook.

Student teams are encouraged to create their own experimental plans and assign roles to team members. It is recommended that one role should be 'library research'. Student plans are vetted by the entire class for relevance to the course. If deemed worthwhile, student teams are allowed to bid for three or four 4-hour laboratory sessions. This allows them time to set up and run their experiment(s). Since the experiment is part of the teaching programme of the course, students are required to bring their findings back to the class as a formal presentation. Since many experiments take several hours to set up and, usually, a few minutes to run, students are encouraged to use video to record their experiences.

Aims:

- Principal objective is demonstrating concepts — allowing observation of geomorphic processes

- The practical is particularly appropriate for demonstrating non-linearity

- The skills content is about designing and running an experiment

- Learning to work as a team and time-management are formally encouraged by role setting

Learning objectives:

Students will:

1) gain understanding of the character of erosion processes and the problems of erosion control technologies by direct experience (one of the most important understandings gained is that erosion rates are governed by exponential laws)

2) learn how to design and execute their own 'control' experiment or suite of experiments

3) reinforce prior learning of team-working skills

4) gain further experience of managing the logistics of problem-solving within strict time and resource constraints

Techniques included:

- Planning how to use a single piece of heavy-engineering equipment, and how to model the natural geomorphic processes in a controlled experiment

- Running a series of experiments with different parameters

Management of laboratory access:

The flume is housed in its own small (listed) building, and access is very restricted. A small group of students and the technician is all that can fit in. The flume is demonstrated for the students for one session, and then they book in times more flexibly to run their own experiments. As a consequence of health and safety issues, the flume is only run with a technician present.

Assessment:

The student group produces an initial plan, in the form of an oral presentation, to the whole class, after seeing the flume demonstrated. This is 5% of the module assessment. The main write-up is a group report in the format of a scientific paper, with references. This is 20% of the assessment. The group also give an oral presentation, with videos if available, to the whole class, accompanied by a poster. This is 15%. The module also includes weekly class tests on the lecture content. Although these are only awarded 2% each they are popular with the students and maintain attendance and revision of the conceptual part of the course.

Comment:

This is an example of effective use of a major piece of heavy engineering with limited access, by setting group tasks which can be run with the help of a technician. The use of video to share the outcomes (which are dramatic and short-lived) allows the benefits of the experimental approach to reach the whole group. The opportunity for students to witness geomorphology in action is highly motivating.

Case study 8:   Images of the Third World

Innovator: Margaret Harrison, Cheltenham and Gloucester College of
                          Higher Education

Practical: Various

Level and numbers: Level 2, 45 students

Preparation: Most students have followed initial human geography modules

Context:

This complete twelve-week module in human geography has a variety of imaginative practical sessions. Questioning and reflection are encouraged throughout the course. Some practicals comprise the full three-hour period of contact time, others follow a lecture. The module begins with an 'Any Questions' session which explores students' preconceptions. In the first week student groups set questions. In the second week they are answered by an invited panel — sometimes this includes experts, but usually it involves informed speakers who adopt a role for the event. The practical for the third week is concerned with numerical analyses of socio-economic data, but these data are selected to challenge common images of the developing world, and the practical includes a feedback session which allows this to be explored.

Week four comprises a closely-structured role play on the theme of imperialism. Three settings are provided in some detail: Senegal under French colonial power in the late 1950s;

an issue of land conflict in India under the Raj; and obtaining a Green card for the USA in the present day. Following a lecture on colonialism — which looks at utopian views of the colonists of South America — students are given the cast details and running order of events. Selected role plays are re-run for the benefit of the whole class, and this is followed up with discussion.

The next practical is concerned with depiction of the third world in posters, newspapers and maps. Group activities are structured around critical analysis of these media (Harrison, 1995). In the subsequent week students watch a full-length feature film, either based on the Caribbean in colonial times or modern West Africa, and they complete worksheets designed to encourage reflection. In week seven the practical involves listening to music of the third world, and an invited speaker lectures on themes such as the appeal of non-western music in the nineteenth century.

On the basis of set reading in week eight, student groups prepare a presentation on agricultural issues in sub-Saharan Africa and deliver these oral presentations in the practical session. In the following week there is a conventional seminar, and in week ten a "game" is played on the theme of seeking and keeping work. Students allocate 'worker' role cards, four from the formal sector, four from the informal sector. They pause to draw a diagram as a group indicating their likely interrelationships, and then each individual outlines the way they spend time and their income and outgoings. On the throw of a die the group is placed in either an economic recession or a boom — with some economic details provided. The 'game' requires them to manage their time and finances within this given economic context, and to examine the effect on their interrelationships. The last practical examines the issue of tourism, and is either a debate about banning tourism, or the deconstruction of advertisements and brochures.

Aims:

All the practicals are fully integrated into the module and designed to illustrate and reinforce concepts. Transferable skills of groupwork, decision-making, developing arguments, summarising material and oral presentation are all involved. Examining representations — both other's and their own — is explicitly developed. The module as a whole leads to a real development of students in this area of reflection.

Learning objective:

When the students have completed these practicals they will have been challenged to question their own motives and reasons for studying "What is the Third World".

Assessment:

50% examination, with one compulsory sectionalised question related specifically to practical work; 50% coursework which is writing a script for a documentary. The latter should demonstrate their ability to reflect and to examine issues and perceptions more deeply.

Comment:

This case study provides a model of variety of practical exercises in terms of styles of teaching and learning, but with the common aim of developing students' awareness of representations.

Case study 9:   Weathering, rainsplash and rivers

Innovator:	Callum Firth, Brunel University
Practical:	Research experiments in geomorphology (in groups)
Level and numbers:	Level 2, 30 students
Preparation:	Level 1, some geochemistry and particle size work, no experiments.
	Introductory lecture covers general experimental design, the equipment is demonstrated, and each group has a tutorial session with teaching staff for feedback on their ideas and design, before they begin.

Context:

Within a module on *Global Geomorphology* student groups choose one semi-structured experiment out of three (freeze-thaw, rainsplash erosion, channel development). Lectures on processes are run alongside the practicals. Each experiment simulates geomorphic processes using simple materials and equipment. For the simulation of freeze-thaw weathering, students are provided with a domestic freezer, cylinders of sandstone, chalk and granite (prepared in the rock-cutting lab.) and four specified conditions: outside the freezer, dry and exposed to freezing cycles, wet and exposed to freezing cycles, and wet and saline and exposed to freezing cycles. Students choose the frequency of cycle, the number of cycles, and the concentration of salt. It works well, with differences between rock types, and also forms of disintegration. However the emphasis is on the experimental design, and the use of literature, rather than the exact outcome.

The channel development experiment operates within a narrow flume which can be set at various angles and used with different sediment types. The rainsplash experiment is more open for student design. They have to model sediment on a slope at a particular angle and then decide how to deliver rain of a chosen intensity (for example, by pipette, meshes to drip, slow spinning through a fan, hose and sprinkler). The sediment or soil is usually placed on large diameter filter paper, marked 'upslope' and 'downslope' and the amount of sediment moved by rainsplash is recorded by weighing these sectors.

There is a follow-up tutorial with teaching staff, to examine the data, and consider progress.

Aims:

- Demonstrating geomorphological processes in operation (often cannot be done in the field)
- Experience of experimental design and problem solving.

Techniques included:

- Experimental design, scientific report-writing

Management of laboratory access:

Students are introduced to the materials in an initial 45 minute session following the introductory lecture. They then book in with the support staff and laboratory to conduct the experiments over a five week period.

Assessment:

Written up individually as a scientific report. Students are given guidelines for writing, and the report is expected to include references to literature and statistical analysis if the data are appropriate. This counts as 25% of the module, with 25% on fieldwork and 50% on the exam.

Comment:

An interesting semi-structured approach to compare with Case Studies 7 and 10, also concerned with geomorphic processes. The course includes opportunities for induction sessions which involve tutorials as well as demonstration of equipment. Follow up involves another tutorial, giving some structured opportunity for reflection.

Case study 10:   Stress, strain and plasticity

Innovator:	Brian Whalley, Queens University, Belfast
Practical:	Experiments on properties of materials
Level and numbers:	Level 2, 35 students
Preparation:	Level 1, a basic physical geography course, which includes some practicals. However, students have very mixed science backgrounds.

Context:

Within a module *Introduction to Geomorphology* students, in groups of three, are set a series of open-ended tasks. The materials are provided to give experiential learning in some physical principles which are directly related to geomorphic processes. Lectures take place alongside, and there is fieldwork earlier in the course. In the first experiment students are provided with 'slopes' which can be set at different angles and provided with different roughnesses (gravel, sandpaper). They are also provided with cuboids of various rock types, differing in mass and shape. Students have to structure their own exploration and experiments in which they investigate the various circumstances when sliding (or toppling) occurs.

In the second experiment the groups are provided with pieces of wooden dowelling, 30 cm long. Some are round and some are square in cross section. A dowel is clamped at one end and weights are suspended at the other, whilst deformation is recorded and the mass at point of failure is noted. Cross section shape is less important than area (deliberately misleading!). The third experiment involves a cuboid of plasticene. Weights are placed on top and deformation measured. The plasticene is also warmed, and the experiment repeated.

Aims:

- Demonstrating concepts of stress, strain and elasticity
- Assisting students to think of properties of materials in a theoretical way

- Developing problem solving and experimental design skills
- Setting and testing hypotheses

Learning objectives:

When students have completed this practical they will be able to:

1) think through a simple practical problem

2) relate theory to practical experience

3) provide 'tests' of theory

Techniques included:

- Experimental design

Management of laboratory access:

Students can book into the laboratories for as long as they need. In practice they use one of three timetabled 2 hour practicals and about 1 hour more of their own time for writing up. All the experiments are carried out in the absence of teaching staff, deliberately. Support staff are around.

Assessment:

Written up individually as a scientific report. This comprises one third of the coursework on this module. Coursework is 40% of the total assessment.

Comment:

This practical involves some very open enquiry learning which is distinctive. It is successful once students have found their feet. It may help that these students have a common geography background which included some practical work at Level I, and they have already been on fieldwork together earlier in this course. As a consequence they are not strangers, and discovery learning which is relatively unstructured is less likely to promote a hostile reaction, and, in those circumstances, likely to be very effective as a teaching tool.

Case study 11:   Projects on geographical issues

Innovators: Mick Healey[1], Ian Livingstone[2], Peter Vujakovic[3], Ian Foster; Coventry University
now at [1]Cheltenham & Gloucester College of Higher Education; [2]Nene — University College Northampton; [3]Canterbury Christ Church College

Practical: Four projects to replace first year practicals in geographical skills

Level and numbers: Level 1, 120-200 students

Preparation: Students have mixed backgrounds, including some with no geography.

Context:

The projects were devised as a new way of delivering practical skills at the foundation level (Healey, 1992). The intention was to replace a techniques-driven programme, which was designed around individual student effort and assessment, with an issues focus and collaborative groupwork. The coverage of practical techniques was reduced, omitting surveying and field techniques, and concentrating on the choice and application of statistical and computing techniques, and on the wider range of skills involved in presentation and communication.

The four projects were (in chronological order): 'The Social and Economic Geography of Coventry', 'A Third World Atlas', 'Environmental Hazards', and 'Acid Rain in Coventry'. Each was allocated 5 weeks of 2-hour classes. The first project introduced most of the skills. Using the Small Area Statistics of the Census of Population, each group worked with one ward and produced indicators, maps and graphs. They also showed cross-city variations. For the second project the group (now re-formed) chose an area and a theme for an atlas and produced two relevant maps (for example, an economic atlas of southern Africa). The third project used library research to prepare a report on a chosen hazard. Group members took on roles of particular specialists in their research, and put together a joint oral presentation. The last project used a computerised database and required students to perform statistical analysis and to write a 2000 word report.

Learning objectives:

By the end of this course students should:

1) be more aware of sources of geographical information

2) have increased their ability to chose and use appropriate statistical techniques to analyse data

3) have developed their ability to present information in a range of different ways

4) gained useful experience of working co-operatively in groups

Delivery of practicals:

Guidance in the techniques and skills was provided for the students by means of a resource pack, which students would access as needed, in place of formal delivery. The role of the tutors was perceived to be as facilitators rather than instructors. Tutors monitored the activities of the groups by observing and questioning. Tutors were provided with a staff handbook which included guidance on the teaching, learning and assessment of the projects, and suggested how they could facilitate the student learning experience.

Assessment:

Posters, written reports, the maps and oral presentation were elements of assessed work for the different projects. Peer and tutor assessment was included.

Comment:

An example of replacing the 'technique-a-week' format of a Level I practical skills course with issues-based projects and a focus on transferable skills. Assessment was carefully designed to reflect these learning objectives and student evaluation showed that the teaching of statistical and computing techniques in this format was markedly more successful than in the old course. This course received a commendation from The Royal Society of Arts Partnership awards for Innovation in Teaching and Learning in Higher Education, BP Geography prize (1992).

Case study 12: Understanding the electromagnetic spectrum

Innovator:	Ted Milton, Southampton University
Practical:	Making a slide-rule to aid understanding and revision
Level and numbers:	Level 2, 80 students
Preparation:	Students are drawn from a very wide range of backgrounds which includes geographers, about 10% of the total having no science background, and up to 50% non-geographers, including physicists.
	There is no common prerequisite.

Context:

The inter-disciplinary module on *Remote Sensing for Earth Observation* has been offered since 1983, and contributes to a 'Physics with Space Science' degree as well as to 'Geography'. Early in the course it is important to communicate the concept of the electromagnetic spectrum and the consequence of the interaction between electromagnetic radiation and the Earth's surface. Later in the course the students experience more conventional practicals in image-processing. The 'slide rule' is an A5-sized learning aid which each student develops for themselves. Movement of the inner card on side A allows a visual display of the link between a particular wave length, the spectral reflectance of different ground surfaces, and the ability of satellite sensors to measure it. On side B the whole electromagnetic spectrum is shown, with the details of relevant platforms and sensors alongside, and a variety of up-dateable characteristics of each platform are displayed by the sliding inner card.

Students are provided with a pre-printed outline card, and reference material, and have to elaborate on the basic design. Experience has shown that cross-discipline groups are very effective, as geographers bring environmental expertise and physicists have a better initial grasp of the physics. For the purposes of the course these skills need to be integrated, and the activity of making the slide rule assists both socially and intellectually. Cards are usually added to throughout the course, and serve as revision aids. The innovation developed from a computer-based hypertext program. Students benefited from the interactivity of the program but wanted a more portable output — especially since much of the information they gather to put on it comes from the Internet now. The slide rule is interactive, creative and updating involves virtually no expense.

The slide rule was one of three innovations in this course which contributed to Ted Milton's receipt of The Royal Society of Arts Partnership awards for Innovation in Teaching and Learning in Higher Education, BP Geography prize in 1994.

Learning Objectives:

1) Demonstrating concepts and reinforcing theory.

2) Assisting in the integration of physical and environmental concepts.

3) Improving interpretation of graphical, textual and numeric information.

4) Assisting in effective cross-disciplinary student group work.

Assessment:

The slide rule itself is submitted as part of the coursework assessment. It comprises 5% to 10% of the course assessment, of which 50% is exam-based.

Reference:

Milton, E.J. (1994) A folded card slide-rule on the electromagnetic spectrum, *International Journal of Remote Sensing*, 15, pp.1141-1147

Comment:

A simple idea which developed into a great success, appreciated by the students and assisting in group development across subject areas as intended.

Case study 13:   Predicting a volcanic eruption

Innovator:	Phil Gravestock, Cheltenham & Gloucester College of Higher Education
Practical:	Group work 'role play' exercise in hazard management
Level and numbers:	Level 2, generally a maximum of 50 students
Preparation:	An introductory lecture introducing the main monitoring techniques, e.g. COSPEC, seismic studies, tilt measurements

Context:

The practical is based around the activity of four fictitious volcanoes. The student group is, ideally, split into four groups so that each group considers one volcano: if larger groups are involved one or two of the volcanoes could be assessed by more than one group of students. A series of activity reports are available for each of the volcanoes. These activity reports may be daily, or may skip a couple of weeks depending upon the activity of the volcano.

Aims:

To introduce concepts which are difficult for students to appreciate such as the length of time involved during a volcanic crisis, from the onset of precursor activity to the actual eruption, the levels of personal stress involved, and hazard response procedures.

Learning Objectives:

After completing this practical a student should:

1)   have gained experience of working as a team member in a 'stressful' situation;

2)   understand the nature of volcano alert levels and how they are implemented;

3)   appreciate how such decisions affect the lives of people living near a volcano;

4)   realise the importance of baseline measurements and monitoring procedures;

5)   be aware of the timescale involved between volcanic precursory activity and final eruption.

Delivery of Practicals:

Each student group is given an initial briefing paper describing the monitoring techniques available for their volcano, and whether or not a hazard zonation map is present. They also receive their first activity report. In order to receive the next report the students are required to assign an alert level to the situation and consider whether it is necessary to evacuate any of the local towns/villages. Each volcano differs in the amount of information which is supplied (i.e. not all have a hazard zonation map, some do not have SO_2 emission data). The examples are all based, loosely, around actual eruptions, but have been disguised so that students cannot anticipate what might happen. One volcano does not erupt, despite reasonable precursor activity. In order to reproduce the 'stress' levels involved the students have a limited time in which to make a decision.

Assessment:

This practical is not formally assessed. However, each group makes a brief presentation at the end of the session explaining the decisions that they made and how, with hindsight, these decisions could have been improved. This enables a process of reflection.

> **Comment:**
>
> The use of role play to illustrate physical environmental processes, and human attempts to manage those processes, is unusual. It is highly motivating for the students and succeeds in focusing attention on real-world decision making in an imperfectly understood environment.

9.2   Where do good practicals spring from, and what sort of future do they have?

In relation to the case studies included here, the innovators were asked how long they had been running the practical, what it had developed from, what the climate for innovation and sharing good practice was like in their department, and whether they perceived future constraints or problems.

How long does a good practical take to evolve?

All case studies had a good track record, and had been evolving over several years. Innovators drew on their experience as demonstrators in other departments (when doing their PhD), feedback from students, discussions with colleagues about what was 'needed', and, in one case, a need to cut assessment; in that case the previous course had been a standard laboratory skills course with every week assessed, and the new course was focused on a particular problem and included group assessment.

What is the role of the Department and Staff Development?

In every case the innovator acknowledged a good climate for developments, but noted that this was informal (only relating to formal staff development by way of induction courses for new lecturers, who then added to the pool of interested staff) and usually depended on a sub-group of lecturers within the department — sometimes quite a small sub-group. In one case the climate was perceived to have changed adversely between the TQA visit and the RAE assessment and the potential for further innovation was questioned.

Who innovates?

The innovators appeared as a determined and committed set of individuals. In three cases the opposition or scepticism of other members of their department was mentioned (unsolicited!). Most innovators had benefited from the support of their Head of Department.

and how?

Potential problems had been surmounted by a variety of imaginative solutions. Timetabling into the laboratories should have been a major difficulty with increasing group sizes, but by 'blocking' laboratory time and enabling students to have some freedom to book into sessions depending on their other course commitments, systems have been evolved which academic and support staff, and students, are satisfied with (see, for example, details of technical support at one department in Section 6.2). Competing demands on laboratories, such as postgraduate students on taught courses or research, or other subject areas, are also blocked. The 'long, thin' format with practicals every week for twelve or fourteen weeks may be a thing of the past. The other factor in common was good technical support (where it was relevant) and innovators had found technicians willing to support increased 'independent' learning by students in the laboratory or IT area.

Innovation in laboratory-based practicals needs:

- familiarity with current practice in industry and in academic research

- willingness to compromise and adapt that practice to enable a large number of students to obtain that experience (and to overcome structural difficulties of timetables and accommodation)

- awareness of the purpose of the practicals in a teaching and learning context and an ability to design student experience to reach specified learning objectives

10 Alternatives to laboratory work

Whilst the purpose of this Guide is to address issues related to laboratory and practical work, it is our view that careful consideration of the aims and learning objectives of the specific course, module or unit may result, in some instances, in the introduction of alternatives to laboratory work. Indeed a number of authors make reference to the possibility, and value, of substituting laboratory work and practicals with other teaching and learning approaches, for example,

> *"Computer simulations can also effectively reach many of the objectives of laboratory teaching"*
>
> *(Kozma, 1982 cited in McKeachie, 1994 p.137)*

and Holliday (1973, p.41) poses the question,

> *"Is it not possible to design some simulated experiments, in which the student must still make decisions, select data, interpret such things as spectra, correlate observations and comment on results — all without requiring purely experimental skills? Participation in such 'experiments' would enable the student at least to appreciate and evaluate the role of practical work"*

Elton (1973) suggested simulated experiments using film loops or tape/slide presentation as one variation on his 'self-service experiments', in which students undertook independent laboratory work in 'teaching booths'.

Although the two suggestions above were made over twenty years ago, they become of particular interest now since the various media involved in the substitute practicals can, in theory, all be presented on one computer. Moreover, interactivity can be built into a multimedia program, allowing students to place dots on graphs and complete data tables, and even have some feedback about their accuracy. To date, a program providing an effective virtual reality environment for students to be able to undertake the sort of laboratory work described in this Guide does not exist, but a 'Virtual Psychology Laboratory' (http://www.cf.ac.uk/uwcc/psych/stevensonwc/vp-lab) has already been set up on the World Wide Web. This preserves and makes available psychological experiments and their software environment.

Computer assisted learning projects have encouraged experimentation with multimedia delivery of virtual practical and fieldwork experiences.

CAL

One example, now integrated by Jacky Birnie into a Level II course on *Ecosystems* at Cheltenham & Gloucester College of Higher Education, comprises computer assisted learning materials on soil survey, developed at Aberdeen (Dawson *et al.*, 1995). Data on landscape (OS map and aerial photographs), soil profiles, and detailed analyses of soil samples from those profiles are all available and the tutor then designs a practical which involves sampling, analysis and interpretation of results, and writing up. Projects can be enquiry-based — the provision of information is sufficient for students to be able to do a considerable amount of discovery — or more tightly structured. Problem-solving projects can even be constrained by a realistic financial budget which is built into the program.

This computer-based substitute for both field and laboratory work has some advantages. If the program is good, the quality of the experience the students have can be very realistic, but it could also be designed to show something specific — illustrating text book theory with text book practicals. Whether it really prepares individual students for what they may find in independent project or dissertation work in the real world may be questionable, yet it effectively introduces all students to the results of successful field and laboratory work — there is no risk of their experience being clouded by mistakes or teamwork problems.

Developments on the Internet have also offered new alternatives to practicals. Within the 'Virtual Geography Department' (http://www.utexas.edu/depts/grg/virtdep/contents.html) an initiative started at the University of Austin, Texas there is, for example, a page of learning materials in the *Earth's Environment and Society* section. This includes several active learning modules, and case studies on the Galapagos, Sierra Nevada, High Plains aquifer and 'Today's Earthquake Activity'. The latter is an online module, in other words, student activities are included.

'Today's Earthquake Activity'

Recent earthquake data is provided in three tables — accessed by links from the page

Directions for student activity:

- Print out this **blank world map** to use
- Keep a record of the location of earthquakes for an extended period
- Plot the location of earthquakes on your map
- Colour code your marks to indicate relative magnitude
- Use your earthquake locations to infer the location of plate boundaries
- Sketch them on your map
- Compare sketch boundaries with the **actual tectonic plate boundaries**

Also within the Virtual Geography Department are some laboratory exercises, including a 'Cloud Morphology Exercise' at Iowa State University. Photographs of cloud types are sorted by the students into groups, and this leads into further study of the causes of particular cloud types. The frontal situation and the satellite image associated with a particular cloud image can be called up for study. This demonstrates how powerful computer-based exercise can be when such a wide data source becomes available to the individual student.

A 'Global Warming Project' homepage (http://www.covis.nwu.edu/Geosciences/ projects/GLOBAL_WARMING/GLOBAL_WARMING.html) offers a series of techniques that might be used by teachers in their delivery of Level I material on this issue. These include suggestions for spreadsheet modelling of responses to changing energy levels which could be incorporated into geography practicals, and demonstrate the utility of spreadsheet modelling.

Computer simulation

Gibbs (1988) presents an example of use of computer-based simulation of the physiology of respiration in biology teaching. One advantage is that the computer performs complex calculations of vital responses to changes in parameters very rapidly, so students can repeat experiments quickly and see what the outcome is when they adjust circumstances and 'run' the 'patient' again. However, Gibbs notes that trial and error would never succeed in reviving an ailing patient — because the number of biochemical reactions and variables involved is so large. Instead of 'discovery' learning (see Section 3), students must use the literature to come up with hypotheses about what is happening, and focus their experiments on those, more structured, investigations.

In fact, the larger (and more realistic) the data sets made available on computers the more that student practical work will need to be structured to ensure that the exercise is effective in achieving its objectives. Although computers may be seen as a cheap substitute for laboratory experiences, there is a major cost in the time invested by staff in setting up effective independent practical exercises.

This is not the place to debate all the 'pros and cons' of substitute laboratory work — see also discussion of virtual field trips in the 'Fieldwork & Dissertation in Geography' Guide in this series (Livingstone *et al.*, 1998) — but any such alternative to practical work can be assessed against the same checklists of criteria, and the learning objectives and effectiveness of the exercise in achieving them can be established in exactly the way they would be for the practical itself. There may be advantages in the quality of training in statistical techniques, and disadvantages in the lack of groupwork and the unreality of the options for problem-solving or hypothesis testing.

11 Conclusion

In this Guide we have sought to explore and discuss the value of practicals and laboratory work in relation to promoting the quality of student learning in geography within the current and foreseeable social and economic climate of higher education.

The main contributions to the Guide have come from geography lecturers, many of whom have developed their own approaches to practical and laboratory work in isolation.

We see the key benefits to be derived from this Guide as:

1) sharing practical curriculum innovation and development across the discipline

2) providing a framework of issues to be considered and addressed in relation to practical work in geography

In summary, those issues to be directly addressed *by the lecturer* are:

- clarification of the specific aims and objectives of practical and laboratory work

- promotion of active student learning, including the opportunity for reflection

- being explicit about the particular skills to be developed

- consideration of a range of appropriate assessment tasks

a *discipline team or department* could engage in the following:

- staff development and involvement of technical staff and postgraduate demonstrators

- providing appropriate student induction arrangements for laboratory and practical work (including aspects of health and safety)

- making the most appropriate use of lab space and taking up any new-build opportunities to rethink laboratory design

- consideration of alternatives to laboratory and practical work

Whilst we would encourage individual geography lecturers to explore these issues, we would recommend that they be discussed in a wider forum to increase the probability of collective strategic shifts to improve student learning through practical and laboratory work.

Finally, you are requested to contribute to the GDN Web pages (http://www.chelt.ac.uk/gdn), with developments in laboratory and practical work which you, or your department, have undertaken. We are well aware of the United Kingdom dimension of our case studies and references and would therefore particularly welcome international contributions. The process of dissemination of innovative and effective practice is intended to continue.

12 References

Bligh, D., Jaques, D. & Warren Piper, D. (1980) *Methods and Techniques of Teaching in Post-secondary Education* (Paris: UNESCO).

Boud, D., Dunn, J. & Hegarty-Hazel, E. (1986) *Teaching in Laboratories* (Guildford: Society for Research into Higher Education).

Bradford, M. & O'Connell, C. (1998) *Assessment in Geography* (Cheltenham: Geography Discipline Network, CGCHE).

Brown, G. & Pendlebury, M. (1992) *Assessing Active Learning*, Effective Teaching and Learning in Higher Education Module 11 (Sheffield: CVCP Universities' Staff Development and Training Unit).

Exley, K. & Moore, I. (1993) *Innovations in Science Teaching* (Birmingham: Standing Conference on Educational Development Paper 74).

Chalkley, B. & Harwood, J. (1998) *Transferable Skills and Work-based Learning in Geography* (Cheltenham: Geography Discipline Network, CGCHE).

Coats, M. (1991) *Open Teaching Toolkit: learning how to learn* (Milton Keynes: Open University Press).

Dawson, B., Dent, D., Davisdon, D., Nortcliff, S. & Fitzpatrick, E.A. (1995) Soil Surveyor, in: S.B. Heath (Ed.) *MERTaL(TM) Courseware* (Aberdeen: University of Aberdeen).

Denicolo, P., Entwistle, N. & Hounsell, D. (1992) *What is Active Learning?*, Effective Teaching and Learning in Higher Education Module 1, parts 1 & 2 (Sheffield: CVCP Universities' Staff Development and Training Unit).

Elton, L.R.B. (1973) Instructional methods in tertiary science education, in: Billing D.E. & Furniss B.S. (Eds.) *Aims, Methods and Assessment in Advanced Science Education*, pp.69-74 (London: Heyden & Son Ltd.).

Gibbs, G., Gregory, R. & Moore, I. (1997) *Teaching More Students 7: Labs and Practicals with more students and fewer resources* (Oxford: Oxford Centre for Staff Development. Oxford Brookes University).

Gibbs, G. (1988) *Learning by Doing* (London: Further Education Unit).

Guy, J.J. (1982) What's wrong with university chemistry?, *Chemistry in Britain*, 18, p.44.

Habeshaw S., Gibbs G. & Habeshaw T. (1993) *53 Interesting Ways to Assess your Students* (Bristol: Technical & Educational Services).

Haigh, M.J. & Kilmartin, M.P. (1987) Teaching soil conservation in the laboratory using the "Bank Erosion Channel" flume, *Journal of Geography in Higher Education*, 12(2), pp.161-167.

Harper, J. (1996) Laboratory Work: An Integrated Approach to Learning and Assessment in Science Courses, in: Hounsell, D., McCulloch, M. & Scott, M. (Eds.) *The ASSHE Inventory, Changing Assessment Practices in Scottish Higher Education*, p.64 (Edinburgh: Centre for Teaching, Learning and Assessment, University of Edinburgh and Napier University, Edinburgh in Association with UCoSDA).

Harrison, M.E. (1995) Images of the Third World: teaching a geography of the Third World *Journal of Geography in Higher Education*, 19(3), pp.285-297.

Healey, M. (1992) Curriculum Development and 'Enterprise': group work, resource-based learning and the incorporation of transferable skills into a first year practical course, *Journal of Geography in Higher Education*, 16(1), pp.7-19.

Healey, M. (1998) *Resource-based Learning in Geography* (Cheltenham: Geography Discipline Network, CGCHE).

Holliday, A.K. (1973) Curriculum development in tertiary chemistry courses, in: Billing D.E. & Furniss B.S. (Eds.) *Aims, Methods and Assessment in Advanced Science Education*, pp.39-43 (London: Heyden & Son Ltd.).

Horobin, R., Andersen, B. & Williams, M. (1992) *Active Learning in Practical Classes*, Effective Teaching and Learning in Higher Education Module 6, parts 1 & 2 (Sheffield: CVCP Universities' Staff Development and Training Unit).

HSE (1990) *COSHH Guidance for Universities, Polytechnics and Colleges of Further and Higher Education* (HMSO).

Jenkins, A. (1998) *Curriculum Design in Geography* (Cheltenham: Geography Discipline Network, CGCHE).

Kneale, P. (1996) Organising student-centred group fieldwork and presentations, *Journal of Geography in Higher Education*, 20(1), pp.65-74.

Livingstone, I., Matthews, H. & Castley, A. (1998) *Fieldwork and Dissertations in Geography* (Cheltenham: Geography Discipline Network, CGCHE).

Milton, E.J. (1994) A folded card slide-rule on the electromagnetic spectrum, *International Journal of Remote Sensing*, 15, pp.1141-1147.

McKeachie, W.J. (1994) *Teaching Tips: Strategies, Research and Theory for College and University Teachers*. (9th ed) (Lexington Massachusetts: DC Heath & Co.).

Ramsden, P. (1992) *Learning to Teach in Higher Education* (Routledge).

Shulman, L.S. & Tamir, P. (1973) Reseach on teaching in the natural sciences, in: R.N.W. Travers (Ed.) *Second Handbook of Research on Teaching*, pp.1098-1148 (Chicago: Rand McNally).

Smith, J.M. (1984) Experiments relating to the fracture of bedrock at the ice rock interface, *Journal of Glaciology*, 30, 123-125.

Solomon, J. (1980) *Teaching Children in the Laboratory* (London: Croom Helm).

Stark, M.J.R. (1996) Structured reports from undergraduate practical classes, in: Hounsell, D., McCulloch, M. & Scott, M. (Eds.) *The ASSHE Inventory, Changing Assessment Practices in Scottish Higher Education*, p.130 (Edinburgh: Centre for Teaching, Learning and Assessment, University of Edinburgh and Napier University, Edinburgh in Association with UCoSDA).

Sutton, C. (1992) *Words, Science and Learning* (Milton Keynes: Open University Press).

Tamir, P. (1977) How are laboratories used? *Journal of Research in Science Teaching*, 14, pp.311-316.

Ward, A., Baume, D. & Baume, C. (1997) *Learning to Teach: Assisting with Laboratory Work and Field Trips* (Oxford: Oxford Centre for Staff Development. Oxford Brookes University).

Woolnough, B. (1994) *Effective Science Teaching* (Milton Keynes: The Open University Press).